SCIENTIFIC AND RELIGIOUS HABITS OF MIND

A Book Series of
Curriculum Studies

William F. Pinar
General Editor

VOLUME 8

PETER LANG
New York • Washington, D.C./Baltimore • Bern
Frankfurt am Main • Berlin • Brussels • Vienna • Oxford

RON GOOD

SCIENTIFIC AND RELIGIOUS HABITS OF MIND

Irreconcilable Tensions
in the Curriculum

PETER LANG
New York • Washington, D.C./Baltimore • Bern
Frankfurt am Main • Berlin • Brussels • Vienna • Oxford

Library of Congress Cataloging-in-Publication Data
Good, Ron.
Scientific and religious habits of mind:
irreconcilable tensions in the curriculum / Ron Good.
p. cm. — (Complicated conversation; v. 8)
Includes bibliographical references and index.
1. Religion and science. 2. Science—Study
and teaching. I. Title. II. Series.
BL240.3.G66   201'.65'071—dc22   2003026737
ISBN 0-8204-7108-9
ISSN 1534-2816

Bibliographic information published by **Die Deutsche Bibliothek**.
Die Deutsche Bibliothek lists this publication in the "Deutsche
Nationalbibliografie"; detailed bibliographic data is available
on the Internet at http://dnb.ddb.de/.

Cover design by Lisa Barfield

© 2005 Peter Lang Publishing, Inc., New York
275 Seventh Avenue, 28th Floor, New York, NY 10001
www.peterlangusa.com

All rights reserved.
Reprint or reproduction, even partially, in all forms such as microfilm,
xerography, microfiche, microcard, and offset strictly prohibited.

This book is dedicated to my grandchildren, Nicole and Kevin, in the hope that they will develop the critical habits of mind needed to function as free thinkers. The willingness to question authority develops to varying degrees in different people, and parents and teachers can be strong role models for children. In this case a grandparent hopes to offer some insight to his grandchildren on the nature and importance of critical, scientific thought. Developing a healthy skepticism to claims one encounters, while at the same time being open to new ideas, is not easy to accomplish, but it is the basis of free inquiry. Many things in society work against the goal of free thought, including certain religious habits of mind imposed on young children before they are able to reflect critically on ideas and events in their lives. In his 1996 book *The Demon-Haunted World: Science as a Candle in the Dark*, physicist Carl Sagan describes many of the ways people can be fooled into thinking there are simple solutions to complex problems. The ease with which many people slide into pseudoscience and superstition is documented throughout history, and in spite of the advances of science in understanding our natural world it continues today. It is possible to develop a scientific outlook and still maintain a spirituality that many think comes only through the rituals of organized religion. It is my hope for my grandchildren that they will learn the importance of free inquiry throughout their lives while at the same time understanding that compassion toward others, the stated goal of most religions, can be achieved without giving up critical, scientific habits of mind.

Table of Contents

ix	Preface
1	Chapter One: **Three Key Scientific Discoveries**
1	♦ Displacing Earth from Its Exalted Position
3	♦ Evolution of Species by Natural Selection
5	♦ Relativity and Quantum Theory
7	Chapter Two: **Darwin, Einstein, and Russell on Science and Religion**
7	♦ Charles Darwin as Explorer and Revolutionary Naturalist
17	♦ Albert Einstein as Revolutionary Physicist, Pacifist, and Free Thinker
21	♦ Bertrand Russell as Philosopher, Mathematician, and Social Critic
25	Chapter Three: **Scientific and Religious Habits of Mind**
25	♦ Scientific Habits of Mind
28	♦ Religious Habits of Mind
30	♦ Points of Conflict
31	♦ How the Mind Works
34	♦ More on Points of Conflict between Religion and Science
36	♦ Conflicting Religious Beliefs: A Short History of God
38	♦ Conflict Resolution?
39	Chapter Four: **Democracy and Science Education**
40	♦ Is Scientific Thought Unnatural?
41	♦ Natural Selection versus Supernatural Creation

42	♦	In God We Trust and God Bless America
44	♦	Darwin's Dangerous Idea
47	♦	Dealing with the Elephant in the Classroom
48	♦	Taking History Seriously
48	♦	Confronting the Elephant Directly
50	♦	Answering FAQs
51	♦	Consilience
53	♦	The Science Wars and Postmodernism
55	♦	Understanding Human Nature
57	♦	Human Nature and Education
58	♦	Unlearning Old Habits of Mind
59	♦	Questioning Common Ideas about Human Nature
62	♦	Where Do We Go Now?
69		SOURCES AND FURTHER READING
79		THREE PAPERS RELATED TO THE THEME OF THIS BOOK:
79	♦	Paper One: Evolution and Creationism: One Long Argument
87	♦	Paper Two: Using a Science Literacy Pretest in a Science Teacher Education Course
93	♦	Paper Three: Will Human Behavioral Genetics/Sexual Orientation Be the Next Target of the Censors of School Science?
103		INDEX

PREFACE

What are scientific and religious habits of mind and how do people settle into these different ways of viewing the world? Are these mind-sets basically compatible or incompatible? Why is it so difficult to achieve widespread scientific literacy in our schools and in society in general? What can be done to improve students' understanding of science, themselves, and their neighbors? These are among the central questions considered throughout this book. During four decades of work in science education these are questions that have occupied much of my time, and they continue to interest me. However, it has become increasingly clear that habits of mind associated with certain religious beliefs inhibit progress toward a sound science education and, more broadly, toward a solid liberal education.

The ideas of Bertrand Russell, Albert Einstein, and Charles Darwin on science, religion, and education have had a strong influence on the development of my own ideas in these areas. Although the ideas of others have made important contributions to the development of my thinking about the nature of science, education, and religion, these three persons stand out. Perhaps most of all, Russell's ideas on critical, skeptical, open-minded thinking and his social activism influenced me during my formative years as a graduate student and beginning university faculty member and have continued to influence my thoughts and actions to the present day. The reader who is familiar with Russell's work will recognize his influence throughout this book.

Questioning authority is not unique to science, but it was a necessary condition for modern science to take hold and make progress in seventeenth-century Europe. In the face of the strong authority of organized religion in Italy and throughout Europe, Galileo Galilei and a few of his contemporaries were willing to question religious dogma by supporting Copernicus's argument that displaced Earth from the center of the universe. Over two centuries later Darwin continued the assault on religious dogma by publishing *On the Origin of Species* in 1859. Humans were not only displaced from the center of things, they were united with the other animals by descent from common ancestors. Religious authority seemed to be in question at the most fundamental level.

Ever since Darwin published *On the Origin of Species*, religious fundamentalists and others opposed to his ideas on evolution of species by natural selection have argued and fought against the teaching of evolution in our schools. National polls and other research show that most U.S. citizens do not understand the scientific theory of evolution and many feel it should not be taught in the school curriculum. What is it about certain religious beliefs that conflict with the scientific theory of evolution and other well-established scientific knowledge? Chapter 1 of this book includes a brief historical look at three key scientific discoveries and how religious authorities reacted to them. The first discovery involves Nicholas Copernicus's sun-centered planetary system and how Galileo and Isaac Newton completed the scientific theory of our mechanical universe. This new worldview displaced Earth and, in the eyes of many religious authorities, humans and their Creator from the exalted center of all things to a less significant place in a very large, dynamic universe. Gravity was sufficient to explain the movements of the planets around our sun and the movement of falling objects on Earth. A supernatural force was no longer needed to explain them. The second discovery is Darwin's insight, following his five-year voyage aboard H.M.S. *Beagle*, that natural selection for certain traits over other traits within a species population is sufficient to account for the evolution of new species. Again, the God hypothesis is shown to be unnecessary to explain the great diversity of life on Earth and the extinction of more than 99 percent of all the species that ever lived here. The emperor's clothes of religious authority were being removed by scientific explanations of the natural world. Finally, the third scientific discovery (more accurately two discoveries), relativity and quantum theory by Einstein, led to modern physics and the search for a *theory of everything* (TOE), a mathematical description of all fundamental forces in the universe. Although relativity and quantum theory have had a less obvious impact on people's lives, including their religious beliefs, than either a sun-centered planetary system or evolution of species via natural selection, they deserve to be included for a number of reasons. These reasons are identified and discussed in Chapter 2.

Chapter 2 looks at various ideas on religion and science held by Darwin, Einstein, and Russell. Darwin and Einstein are considered by many to be the most important scientists of their respective fields, biology and physics, and Russell was clearly one of the most important and influential analytical thinkers and social activists of the twentieth century. Outside academia Russell is less well known than Darwin and Einstein, but his influence in analytical philosophy and the logical foundations of mathematics remains strong. Known for his outspoken stands against violence and for universal civil rights, Russell's *Why I Am Not a Christian* (1957) continues to be a classic work among free thinkers and skeptics. The book's title reflects Russell's willingness to question religious and political authority in the most direct way.

Chapter 3 considers scientific and religious habits of mind and how they seem to conflict in the minds of many people. Do certain religious beliefs interfere with education in the natural sciences? Emphasis here is placed on Darwinian evolutionary theory and the difficulties and conflicts experienced by students and others who believe in a personal God who watches over them and occasionally intervenes in their lives and the world around them. The often-repeated phrases *In God We Trust* and *God Bless America* represent a belief in supernatural intervention, usually favoring those who follow certain church doctrines. The fact that church doctrine varies greatly across the many varieties of organized religion does not seem to bother many believers. Modern scientific conceptions of the mind/brain are presented briefly and the persistence and nature of God beliefs are considered in light of this recent research.

Chapter 4 summarizes key ideas and suggests ways to educate children and young adults that encourage them to be critically receptive to new ideas, weighing the evidence for and against arguments. In *Consilience: The Unity of Knowledge* (1998) Harvard biologist Edward Wilson challenges us to understand the value in basing human activity on stable, scientific knowledge of our environment and ourselves. An education that takes these ideas seriously is discussed in some detail in the final part of this section. Following the sources and further reading are three conference papers by me that are related to the theme of this book. The first paper, "Evolution and Creationism: One Long Argument," describes an attempt by Louisiana lawmakers to pass a resolution declaring the writings of Darwin racist and how I, along with various colleagues in science at Louisiana State University, responded. The second paper, "Using a Science Literacy Pretest in a Science Teacher Education Course," describes research on the level of understanding of evolution, selected physical science ideas, and the nature of science by preservice science teachers. The third paper, "Will Human Behavioral Genetics/Sexual Orientation Be the Next Target for the Censors of School Science?" summarizes a study of four high school biology classes and discusses ways to resist censorship.

Finally, I want to acknowledge the many colleagues and former students who have helped me to think more deeply about the relationship between science and religion. In particular, Catherine Cummins, Mark Hafner, David Kirshner, Michael Matthews, Bill Pinar, Mike Smith, Sherry Southerland, and Jim Wandersee come to mind among those who offered ideas that contributed to my understanding of the nature and complexity of scientific and religious habits of mind. And most important of all, I thank Elaine, my very special wife and close companion for over forty years. Thank you, Elaine, for your love and support in this project and others over the years.

CHAPTER ONE:
THREE KEY SCIENTIFIC DISCOVERIES

DISPLACING EARTH FROM ITS EXALTED POSITION

Four persons are identified here to represent the ideas that led to modern science in Europe. Although others contributed to what has been described as a revolution in thinking about our physical world, Copernicus, Kepler, Galileo, and Newton are seen by most historians and scientists as key figures.

Nicholas Copernicus (1473–1543)

Niklas Kippernigk, known to most of us as Nicholas Copernicus, or simply Copernicus, is credited with replacing Earth by our sun as the center of our planetary system, thereby demoting Earth and its inhabitants to a far less exalted place in the whole scheme of things. Well educated in law, medicine, mathematics, and astronomy, this Polish bishop of Ermland is remembered mainly for his reformulation of the Ptolemaic Earth-centered scheme of the universe. Copernicus's sun-centered planetary system required far fewer epicycles to explain planetary motions, and, additionally, accurate measurements of planets' distances from the sun could be made. However, humans were no longer at the center of all the action, a fact sure to displease religious authorities who felt that God would naturally place His most important creation (Man!) at the center of all things. Reluctant to publish his work, Copernicus waited until near the end of his life (he died on May 24, 1543) before agreeing to send his manuscript off to the printer. For over a century after his death the Roman Catholic Church would persecute those who embraced the blasphemous ideas of Copernicus, with Giordano Bruno and Galileo Galilei being the best-known victims.

Johannes Kepler (1571–1630)

Scientific work by Johannes Kepler refined and extended Copernicus's ideas of a sun-centered planetary system and a very large universe, even though the keepers of holy stories did their best to censor the blasphemous ideas that they thought dishonored their God's wondrous and perfect creation. Kepler was a mathematician driven to discover the divine nature of the planetary orbits, and he used the astronomical data of Mars gathered by Tycho Brahe to conclude that Mars moved not in a circle around the sun but in an elliptical orbit. Although Kepler and others before him had assumed that God would choose the perfect circle as the path for Earth and the other planets, he was forced by the observational data for Mars to realize that its orbit was elliptical.

This discovery was followed by painstaking work over more than ten years resulting in two more very important discoveries by Kepler on the nature of planetary orbits. He found that a line from the sun to a planet sweeps out equal areas in equal times and that the square of the time of the revolution/period of a planet is proportional to the cube of its average distance from the sun. Collectively, these discoveries are now known as Kepler's three laws of planetary motion, and they established the fundamental importance of using observational data rather than assumptions (for example, that God preferred circles for orbital motion) in trying to understand nature. Kepler's spiritual quest to discover the harmonies of the celestial spheres of the universe led him initially to prove the perfect and divine nature of circular planetary orbits, but using real observational data he eventually discovered that it is the less divine elliptical path that is taken by planets in their journey around the sun.

Galileo Galilei (1564–1642)

Galileo Galilei is often credited by scientists and historians with founding the experimental method and modern science, but he is remembered by the general public more for his clashes with the Catholic church. Shortly after the telescope was invented by Dutch scientist Hans Lippershey, Galileo, then a professor of mathematics at Padua University in Italy, made his own telescope in 1609 and began using it in the garden behind his house to view Jupiter, our moon, and other objects in the night sky. He discovered the moon's mountainous surface, Jupiter's moons, and that the Milky Way consists of a great number of individual stars. Each of these discoveries was in sharp disagreement with then-current thinking about the universe that had been in vogue for nearly two thousand years, since the time of Aristotle (b. 384 B.C.), and decreed true by the powerful Catholic church. In 1610 Galileo published his many observations in *The Sidereal Messenger* and soon became well known for his keen observations and willingness to report his ideas in the face of threats from established religious powers. He argued that the Bible should be interpreted as metaphor rather than as literal truth where claims about nature were concerned and that scientific thinking should take the lead in helping us understand our natural world. For this and for his later claims in *The Dialogue on the Two Chief Systems of the World* (1632) that Copernicus's sun-centered planetary system was indeed correct, Galileo was tried in Rome by the Inquisition in 1633 and found guilty of blasphemy and sentenced to house arrest for life. It was one thing to hypothesize that the sun rather than our Earth is at the center of our planetary system, but another to claim it as truth. By recanting his claims and signing a letter of apology, Galileo avoided the fate of Bruno, who was burned at the stake in 1600 for supporting Copernicus's ideas of a sun-centered planetary system.

Galileo's final years (he died on January 8, 1642), though spent under house arrest, were very productive in terms of his research on moving bodies, resulting in his *Discourses on the Two New Sciences* being published in 1638, a year after he lost his eyesight, which had been gradually fading for years. For this work, for his earlier work in astronomy, and perhaps most of all for his willingness to question the authority of the dogma of the powerful Catholic church, Galileo is considered by many scientists to be the father of modern science. His commitment to the use of observation, experimentation, and mathematics in trying to understand nature's laws was extremely important to later scientific inquiry.

Isaac Newton (1642–1727)

Born on December 25 in the same year Galileo died, Isaac Newton invented the mathematics (calculus) needed to fully study the rate of change of motion that Galileo had begun. His investigations on force and motion as a young man resulted in the publication of his hugely important book *The Mathematical Principles of Natural Philosophy* in 1686. What we now call Newton's three laws of motion are contained in this work and they unify scientific thought regarding the various motions of the universe, including motion on the Earth. Until Einstein introduced his relativity/invariance theory in the early twentieth century, Newtonian mechanics was unchallenged as the foundation for all study of motion. Newton's work was the culmination of the work of a number of earlier scientists, including Copernicus, Kepler, and Galileo. Most of these scientists, and perhaps Newton in particular, believed they were discovering God's laws, and their work served to glorify His incredible genius in creating a universe that operated according to such precise laws. By the time Newton published his work in 1686 many religious authorities had begun to realize that they needed to accept the overwhelming evidence of a sun-centered planetary system. This bitter pill was consumed, and it was gradually accepted that scientific inquiry is superior to religious fiat in determining how the physical universe operates. Finally in 1822 the Roman Holy Office of the Catholic church permitted the publication of books that taught what Copernicus had argued nearly three centuries before, that Earth, like the other planets, moves around the sun. Just thirty-seven years later the religious establishment would be in for another affront to their holy authority when a second Englishman came along with the worst news of all: even the living world operates according to natural law! The role of the gods was about to be diminished drastically in the eyes of many.

EVOLUTION OF SPECIES BY NATURAL SELECTION

Charles Darwin (1809–1882)

Tremendous progress in the sciences of physics and chemistry was made during the eighteenth and nineteenth centuries, but the same was not true of biology until Charles Darwin came along. As a young man Darwin took a five-year (1831–36) voyage aboard the British ship *Beagle* as naturalist and companion to the ship's captain, and he began to question the creationist account of life's origins as described in the Biblical book of Genesis. After observing many things about nature and considering the fossil record during his voyage of discovery, he realized two years later that Thomas Malthus's idea of population growth being limited by available food supplies was the key to understanding the evolution of living things. Nature selects those in a given population that are best adapted to survive and pass on their genes to future generations, so the God hypothesis invoking supernatural causes is simply unnecessary. Understanding the social and philosophical implications of his idea of evolution of species by natural selection, Darwin was reluctant to publish his work. Most of his friends and colleagues, like the overwhelming majority of British men and women of his time, believed that species were created separately by the Great Creator. His dislike of personal confrontation and controversy and his love for his religious wife, Emma, and their children caused Darwin to delay publication of his ideas on evolution until another naturalist, Alfred Wallace, developed the same ideas about natural selection as the mechanism for evolution. Their papers on evolution of species by natural selection were presented to the British Geological Society on July 1, 1858, and Darwin's famous book *On the Origin of Species by Means of Natural Selection*, which had been in progress for over a decade, was published in November 1859. The relationship between science and religion would never be the same.

The religious establishment reacted predictably to the publication of Darwin's book. Having reluctantly accepted a sun-centered planetary system, with Earth relegated to a minor position in a huge, dynamic universe, now religious believers were faced with ideas that seemed to further diminish the need for supernatural intervention in the real world. It was simply natural environmental pressures acting on variation within populations of organisms, selecting for those that are better adapted, that explained the great diversity of life and the fossil record. What was left for a supernatural power to do? Creationists then (as they do now) fought to keep what they saw as necessary interpretations of holy books. They said, as now, that Earth is too young (only six thousand to ten thousand years old) to allow for evolution by natural selection, and that such wonderfully complex things as eyes and butterflies, not to mention human beings, must have been designed by an all-powerful supernatural being. Surely the tinkering (natural selection) by Mother Nature cannot explain everything we see! Endless arguments by the creationists continue against evolution via natural selection as described by Darwin and modern science.

Although Darwin's *Origin of Species* is the best known of his many books and other publications, two books published over a decade later speak more directly to the implications of his grand theory for humans and their place in the whole scheme of things. *The Descent of Man, and Selection in Relation to Sex* (1871) and *The Expressions of Emotions in Man and Animals* (1872) present evidence that humans are descendents of "lower" animals and that emotions are the result of the inheritance of behavior from other animals. What he had left unsaid in *Origin* about the implications of his theory of evolution for humans is discussed in detail in *Descent* and *Expressions*. For many religious believers, then and now, the scientific theory of evolution might apply to the animal world but not to God's centerpiece of creation, Man. These believers want to see humans as special exceptions to the laws of nature, especially where human behavior is concerned, and they are willing to go to great lengths to deny the relevance of Darwin's ideas to humans. As we will see later, modern science has extended evolutionary theory to include human behavior and even ethics and moral behavior.

Many other scientists since Darwin have refined and extended his ideas, but for now credit for the key idea of evolution of species by natural selection is placed at his feet, just as credit for the key idea of relativity theory is given to Einstein. Even though genetics has become dominant in the world of biology since the discovery of DNA's structure in 1953, it is evolutionary theory that explains and unifies the many areas of study within and even beyond the biological sciences. Only in the light of modern evolutionary theory do things in biology make sense, as Russian-American geneticist Theodosius Dobzhansky (1900–1975) observed a few years before his death.

RELATIVITY AND QUANTUM THEORY

Albert Einstein (1879–1955)

As a young physicist working in a patent office in Berne, Switzerland, Albert Einstein, in one remarkable year, 1905, had five papers published that were to transform the field of physics. Two of the five papers were on relativity theory, modifying the space-time foundations of Newtonian mechanics to make it more consistent with the new electrodynamics of James Maxwell. As the key person behind the birth of modern physics (relativity and quantum theory), Einstein was to become the best-known scientist of the twentieth century. Although his work had less immediate, obvious impact on organized religion than did Darwin's ideas, as relativity and quantum theory became more fully developed the certainty associated with Newtonian mechanics and organized religion began to change. Traditional conceptions of space and time were challenged by relativity theory, and quantum theory required equally challenging reconceptualizing of fundamental ideas about matter and its interactions.

Science has been characterized as a process of describing the familiar in terms of the unfamiliar (Wolpert, 1992) and it is certainly the case with Einstein's relativity theory. Just as displacing Earth from the center of the universe and replacing supernatural creation with natural selection required *uncommon sense* rather than common sense, relativity theory replaced the concepts of absolute space and time with a space-time continuum. The principle of relativity can be stated as *The laws of physics are the same in all inertial systems*, and Einstein is said to have preferred the term *invariance* rather than *relativity* as more accurately descriptive of his theory. However, *relativity* won out and many people are misled by the more common meaning of this term. Another basic postulate of relativity theory is *The speed of light in space is the same for all systems and observers*. From these two postulates or assumptions Einstein derived his famous theory.

In late 1915 Einstein presented another paper on relativity that extended his earlier work to include systems undergoing acceleration. Known as *general* relativity rather than *special* relativity, he assumed that being in a gravitational field is equivalent to being in an accelerated system. This principle of equivalence predicted, among other things, that light rays are deflected as they pass stars, and four years later, on May 29, 1919, during a total eclipse of the sun, test results supported the prediction. All tests of relativity theory since then have been similarly supportive. From the slowing of clocks and other physical phenomena in gravitational fields to black holes, the various un-commonsense predictions of relativity theory have been confirmed. Newton's laws of mechanics are still valid for most practical applications involving objects we see in our environment, but the concepts of space and time and mass and simultaneity and so on involved in relativity theory are very different. We will see later how these different concepts that require uncommon rather than common sense can affect religious/supernatural ideas.

Chapter Two:
Darwin, Einstein, and Russell on Science and Religion

Each of these great intellects offers a unique perspective on science and religion. Although Bertrand Russell was not a scientist in the same sense as Charles Darwin and Albert Einstein were active scientists throughout their lives, he wrote extensively on these two important forces in society.

Charles Darwin as Explorer and Revolutionary Naturalist

Following the Copernican-Galilean-Newtonian revolution in the seventeenth century, in which Earth was displaced from the center of all things and the universal nature of physical law was established, the next major scientific revolution/epistemological rupture (Bachelard, 1984) occurred in 1859 with the publication of Darwin's *On the Origin of Species*. After Newton it was clear that nature (gravity) needed no assistance from supernatural forces to keep the planets and the stars going their various ways, and after Darwin it was just as clear that nature (natural selection) needed no assistance to maintain the evolution and extinction of species. The role of the gods was diminished considerably after Darwin, although the influence of organized religion, especially in the United States, continues to be a powerful force in many people's lives.

Born on February 12, 1809, in Shrewsbury, England, Charles Robert Darwin gave few indications during his first twenty-five years that he would become the best-known and most influential biologist of his era and, many think, of all time. Had it not been for a long voyage aboard the British sailing ship *Beagle,* where he was able to observe geological and biological phenomena in various locations around the world, it is highly unlikely that Darwin would be a household name today. As a child and young man Darwin loved the outdoors, fishing, hunting, and collecting specimens at every opportunity. It was here that he learned much about careful observation that would be invaluable to him during his voyage on the *Beagle*. While at Cambridge University (1829–31), he was strongly influenced by botanist John Henslow and by his reading of several books, including Alexander von Humboldt's *Personal Narrative* (1818) and John Herschel's *Introduction to the Study of Natural Philosophy* (1830). Following completion of his undergraduate degree in 1831 from Cambridge University, Darwin received an invitation from Captain Robert Fitzroy to join him as naturalist aboard the *Beagle* on what was to be a two- or three-year voyage to survey the coastline of South America. In what turned out to be a nearly five-year (December 27, 1831, to October 2, 1836) adventure, Darwin later described those experiences as by far the most important of his life.

Initial Obstacles

The usual obstacles to conceptual change in science are lack of relevant knowledge and the inability to see things from a new perspective. Both are needed to develop new directions and new ways of understanding nature. When Darwin left England creationist beliefs dominated nearly everyone's thinking, including that of the best scientists. Nobody could understand how new species arose, so the God hypothesis was used to "explain" it. God created things, including all life, in a large burst of activity that many called *catastrophism*. Because no natural mechanism could be imagined at the time to explain the evolution of species, the supernatural mechanism of God was invoked. A further assumption was that humans were God's extra-special creation and, therefore, they were viewed as separate from other species. As a creationist of sorts, Darwin had many obstacles to overcome before he could *see* the now obvious mechanism of natural selection. Most scientific discoveries seem obvious after the fact.

Darwin's many observations of nature during his voyage of discovery positioned him to see the importance of T. R. Malthus's ideas of population being limited by food supply. Without these observations and corresponding changes in his thinking, reading Malthus's 1826 essay on population in 1838 would very likely not have led him to the great idea of natural selection.

Perhaps even more important than the many observations of nature accumulated during the five-year voyage was Darwin's willingness to question and ultimately change his creationist beliefs as natural evidence accumulated. Although he continued to hold certain supernatural beliefs at the end of his sea voyage, Darwin had begun to doubt the creation account in the Bible. His reading of geologist Charles Lyell's *Principles of Geology* (1830–33) during the trip contributed to his questioning of the Bible's account of creation. Unlike other geologists of his time, Lyell argued for uniformitarianism over catastrophism. In other words, the dynamic processes of Earth seen in Lyell's time were assumed to be similar to the processes throughout geologic time. This view of the stability or uniformity of Earth's processes over a very long time was opposed to the catastrophic creationism dogma derived from Biblical interpretations. Leonard Engel, editor of the Natural History Library edition (1962) of *The Voyage of the Beagle*, asserts, "The Lyell book and his [Darwin's] observations of the fossil record nibbled away at Darwin's Creationist beliefs throughout the voyage. By the third year he probably held them no longer" (p. xv). Very early in the voyage Darwin marveled at the existence of great beauty in the vastness of the ocean and in the sea creatures living there with nobody to enjoy it, causing him to question the purpose of it all. The willingness to question the authority of the Bible and instead use observation and reason did not come all at once but accumulated as the five-year voyage continued. Perhaps because Darwin was away from the usual pressures of creationist colleagues and friends in England to make his ideas conform to more "acceptable" notions, it was easier to question the old creationist stories. Left mostly to his own thoughts and the considerable evidence he collected during that five-year period, Darwin was developing the habits of mind needed to work toward his theory of evolution of life by natural selection.

Important Observations

From among the great many experiences and thoughts reported by Darwin in *The Voyage of the Beagle* (1842 edition, published 1962) my colleague Jim Wandersee and I (Good and Wandersee, 1997) chose five that we thought represent critical junctures in the development of his theory of natural selection. We looked for exclamations of surprise, wonder, and astonishment in Darwin's writing as he encountered new phenomena and as he reflected on his observations, treating *The Voyage* as a database to be analyzed for indications of turning points in Darwin's conceptual development. Much of what appears in *The Voyage* was taken directly from his notes recorded daily aboard the *Beagle* or as he paused during overland treks through the countryside. Although his visit to the Galapagos Islands during September and October of 1835 is often referred to as the defining event of Darwin's trip, there were many important experiences and reflections prior to the Galapagos visit.

OBSERVATION *1*: While in Rio de Janeiro in 1832, Darwin compared the relation between plants and insects of the same families across different countries: "We see here in two distant countries a similar relation between plants and insects of the same families, though the species of both are different" (p. 32). This shows that Darwin was always comparing new data to old and considering possible causes of observed patterns and relationships.

OBSERVATION *2*: During the summer of 1833, while in Maldonado, Darwin noticed the variation of vegetation from north to south: "Hence in the southern and northern parts of the continent, the forest and desert lands occupy reversed positions with respect to the Cordillera, and these positions are apparently determined by the direction of the prevalent winds" (p. 47). The gradual variation in the forms of species across space suggested local changes in a common original species, rather than separate creations of new species.

OBSERVATION *3:* Also while in Maldonado during the summer of 1833, Darwin noticed the tendency of mockingbird species to vary gradually across space:

> On the wide uninhabited plains of Patagonia another closely allied species [of mockingbird], O. Patagonia of d'Orbigny, which frequents the valleys clothed with spiny bushes, is a wilder bird, and has a slightly different tone of voice. It appears to me a curious circumstance, as showing the fine shades of difference in habits, that judging from this latter respect alone, when I first saw this second species, I thought it was different from the Maldonado kind. Having afterwards procured a specimen, and comparing the two without particular care, they appeared so very similar, that I changed my opinion; but now Mr. Gould says that they are certainly distinct; a conclusion in conformity with the trifling difference of habit, of which, however, he was not aware (p.54).

So we see that long before reaching the Galapagos, Darwin was sensitive to slight differences in species across habitats, and the mockingbird was prominent among his observations.

OBSERVATION 4: Late in 1833, just two years into the voyage, Darwin is aware of the importance of even minor characteristics of species in affecting their ability to survive in changing environments:

> When the pasture is tolerably long, the niata cattle feed with the tongue and palate as well as common cattle; but during great droughts, when so many animals perish, the niata breed is under a great disadvantage, and would be exterminated if not attended to; for the common cattle, like horses, are able just to keep alive, by browsing with their lips on twigs and reeds; this the niatas cannot so well do, as their lips do not join, and hence they are found to perish before the common cattle. This strikes me as a good illustration of how little we are able to judge from the ordinary habits of life, on what circumstances, occurring only at long intervals, the rarity or extinction of a species may be determined (p. 147).

About two months later (early January 1834) Darwin continued to think about the relation between slight changes in habitat and extinction of species:

> This wonderful relationship in the same continent between the dead and the living, will, I do not doubt, hereafter throw more light on the appearance of organic beings on our earth, and their disappearance from it, than any other class of facts (p. 174).

It is clear that Darwin was thinking of the implications of his observations, preparing himself to realize the importance of Malthus's ideas on the limits of population growth when reading *Essays on Population* in October 1838, two years after he completed his long voyage.

OBSERVATION 5: By the time the *Beagle* arrived at the Galapagos Islands in September 1835, Darwin was well prepared to notice the many interesting and unusual species that existed there. A few of his comments about that stop in his journey show just how important those few weeks were to his thinking:

> Hence, both in space and time, we seem to be brought somewhere near to that great fact—that mystery of mysteries—the first appearance of new beings on this earth (p. 379).

> It was most striking to be surrounded by new birds, new reptiles, new shells, new plants, and yet by innumerable trifling details of structure, and even by the tones of voice and plumage of the birds, to have the temperate plains of Patagonia, or the hot dry deserts of Northern Chile, vividly brought before my eyes (p. 393).
>
> I have not as yet noticed by far the most remarkable feature in the natural history of this archipelago; it is that the different islands to a considerable extent are inhabited by a different set of beings (p. 394).
>
> I have never dreamed that islands, about fifty or sixty miles apart, and most of them in sight of each other, formed of precisely the same rocks, placed under a quite similar climate, rising to a nearly equal height, would have been differently tenanted. (p. 394).
>
> My attention was first thoroughly aroused, by comparing together the numerous specimen, shot by myself and several other parties on board, of the mocking-thrushes, when to my astonishment, I discovered that all those from Charles Island belonged to one species (Mimus trifasciatus); all from Albemarle Island to M. parvulus; and all from James and Chatham Islands (between which two other islands are situated, as connecting links) belonged to M. melanotis (p. 395).

By the time Darwin left the Galapagos Islands (on October 20, 1835), he had amassed enough direct observations to prepare himself to see the importance of Malthus's idea of human populations quickly outstripping their food supply unless their growth is restrained by mechanisms such as famine, disease, and war. In nature, competition for food and related necessities of life (for example, water, soil, sunlight, shelter) within populations of the same species and among different species became the natural selection mechanism in Darwin's theory of evolution of species. Supernatural explanations were no longer necessary to explain changes in organisms over time, and the implications of this discovery would be felt far beyond the world of biology.

From Believer to Unbeliever/Agnostic

Although Darwin was a *believer* when he left on his voyage of discovery in 1831, his religious beliefs were not of the *young Earth* variety that many religious fundamentalists promote today. From the fossil record he understood that Earth was very old, certainly far older than the few thousand years proclaimed by those who claimed to know the true word of God in their Bible, and that a great many species had become extinct. What he and others did not understand was how new species came into being, so they could not argue effectively against supernatural intervention as the explanation for new species. It was curious, however, that an all-knowing God would design species that were destined to become extinct.

Some room for questions like this was provided through the influence of Darwin's agnostic grandfather Erasmus Darwin, who tried to understand evolution in natural rather than supernatural terms. Also, Darwin's physician father, Robert, was an unbeliever, and Charles's older brother, Erasmus, had become an unbeliever by the time Charles completed his great voyage in 1836. So Charles was raised in an atmosphere that tolerated unbelief, and even though he said he started his voyage aboard the *Beagle* as a believer he was influenced by his family to question rather than blindly accept authority. Although he claims in his autobiography, written in 1876, that he did not doubt the *strict and literal truth of every word in the Bible* (Barlow, 1969, p. 57) when he started his voyage, it is clear that he must have been influenced by certain family members and friends who had a tradition of questioning religious dogma. The seeds of doubt and the desire to question and search for causes existed in Charles Darwin as he embarked on his voyage, and the evidence he would amass during the next five years would begin to overwhelm his religious beliefs, which were not strongly held to begin with. However, it seems to be during the two years following his voyage that his questions about religious dogma tipped him from believer to unbeliever, or at least much-weakened believer. Darwin describes how disbelief slowly displaced belief, but the timing is not made clear in his writings:

> Thus disbelief crept over me at a very slow rate, but was at last complete. The rate was so slow that I felt no distress, and have never since doubted even for a single second that my conclusion was correct. I can indeed hardly see how anyone ought to wish Christianity to be true; for if so the plain language of the text seems to show that the men who do not believe, and this would include my Father, Brother, and almost all my best friends, will be everlastingly punished. And this is a damnable doctrine (Barlow, 1969, p. 87).

Both evidence and logic led Darwin to reject creationism and related religious dogma, but he was careful to avoid offending his beloved wife, Emma, who was an orthodox Christian. Following their marriage on January 29, 1839, Darwin tried hard to use caution in expressing his views on religion, knowing it would upset Emma, and this undoubtedly influenced his writings on the topic. Kohn (1989) attributes the ambiguity on religion in Darwin's writings to his fear that Emma would have been devastated if she knew his true thoughts on the subject. The most critical statements in his public writings are found in his *Autobiography*, written in 1876 but not published until 1887, five years after his death.

Some of Darwin's first doubts about his religious beliefs were raised during the years of his voyage when he saw the terrible effects of slavery, as Ernst Mayr (1991) paraphrases: "How could a wise and good Creator permit the unspeakable cruelty and sufferings of slavery? How could he instigate earthquakes and volcanic eruptions that killed thousands or tens of thousands of innocent people?" (p. 13).

So it was not only the biological and geological evidence that influenced Darwin during the voyage, but also the logical analysis of religious doctrine that claimed that God was loving and benevolent but caused horrible suffering and death to innocent people. It was not enough for Darwin to accept the platitude that *God works in strange ways* as justification for the tremendous suffering in the world, with slavery a particularly abhorrent example. And when their beloved daughter Annie died in 1851 at the age of ten, Darwin gave up any remaining belief in the *just and caring God* story that might have remained in the remotest parts of his mind. Although there is little doubt that he was an unbeliever of sorts a decade earlier, after April 24, 1851, Darwin's unbelief was strong and personal.

Following the Voyage

Charles Darwin's ideas on evolution were strongly influenced by the geologist Charles Lyell as well as his own research during the five-year voyage. Against the prevailing thought at the time, Darwin became convinced that evolution, like geological change, was a slow, gradual process. Mayr, one of the most important evolutionary biologists of the twentieth century, points out that "Darwin could have never adopted natural selection as a major theory, even after he had arrived at the principle on a largely empirical basis, if he had not rejected essentialism and physicalism" (Mayr, 1991, pp. 49–50).

Both essentialism and physicalism were closely related to religious doctrine that decreed that all of God's creations were perfect and that nature's laws reflect the precise, physical laws of the Creator. The deterministic spirit of science in Darwin's time was inconsistent with his theory of evolution, in which the laws of chance and statistics are important. The study of nature according to physicalism is the study of God's Grand Design, often called *natural theology*. Darwin gradually saw that natural theology, in which God-given laws are directed toward a final purpose or plan, is inconsistent with the evidence supporting natural selection. How could perfection allow so many species to become extinct, for example? For many millions of years the dinosaurs seemed to be the very embodiment of perfection, as they dominated all life on Earth. Then they quickly disappeared and mammals became dominant among the large animals. To call this progress within God's plan is to use a strange sort of logic.

Following Darwin's return to England in late 1836, but before he had read Malthus's book on population, he recognized that humans were related to other animals by common descent. Although most religious doctrine placed man above and apart from other life forms, Darwin realized that evolution required that humans, like all other life, be linked by common ancestors. He knew that religious authorities would not be pleased to hear that his theory of evolution linked man to other primates, and when *On the Origin of Species* was published in 1859 his fears were realized.

During his time on the *Beagle* Darwin's faith in the accuracy of Christian stories began to weaken, and in the years following the voyage he was very skeptical of most religious doctrine. Once he was free of the restrictions placed on his thinking by religious dogma, Darwin could see that the diversity of life in the world was the result of natural rather than supernatural causes. Rejecting special creation placed Darwin in opposition to the reigning worldview of his day and caused him to withhold publication of his theory of evolution via natural selection for two decades. Emma and many of his friends and colleagues were believers in Christian stories of creation, and Darwin did not want to risk alienating them by publishing his ideas for the world to see and criticize.

In the two decades (1838–58) following his realization that natural selection was the mechanism that explained evolution of species, Darwin accumulated more and more evidence that supported his theory. Had Alfred Russell Wallace not developed a similar theory of natural selection and sent it to Darwin for comments in 1858, it is not clear when Darwin would have published his grand theory. After he received Wallace's manuscript describing natural selection as the mechanism behind evolution, Darwin agreed to have both Wallace's manuscript and his own presented on July 1, 1858, at a meeting of the Linnean Society of London. This event caused him to accelerate the pace of his work on his big species book, as he referred to it, and on November 24, 1859, *On the Origin of the Species by Means of Natural Selection* was published in England. Mayr (1991) describes the importance of that book:

> The impact of the origin was enormous. Quite rightly it has been referred to as "the book that shook the world." In its first year the work sold 3,800 copies and in Darwin's lifetime the British printings alone sold more than 27,000 copies. Several American printings, as well as innumerable translations, also appeared. Nevertheless, only in our lifetime have historians understood how fundamental the influence of this work has been. Every modern discussion of man's future, the population explosion, the struggle for existence, the purpose of man and the universe, and man's place in nature rests on Darwin (p. 7).

Many scientists have made important contributions to their fields, but Darwin's field, evolutionary biology, is unique in a number of important ways. Not only does it serve as the main unifying force for all of biology, its implications go far beyond the field of biology. Perhaps the best statement of the importance and scope of Darwin's work is *Darwin's Dangerous Idea: Evolution and the Meanings of Life* (1995), by philosopher Daniel Dennett. Some of these ideas are covered in the next section in conjunction with a discussion of two of Darwin's books on implications of natural selection for humans.

Darwin's Later Years

With the publication of *Origin* in 1859 Darwin's heretical ideas were finally before the public, and reactions by believers, the vast majority of people then, were predictably negative. Natural selection seemed to make God irrelevant in terms of the day-to-day operations of the living world just as Newton's laws of motion nearly two centuries earlier eliminated the need for a supernatural force guiding the stars and planets and other objects. After the *Origin,* what was left for Darwin to do?

During most of the 1860s Darwin and his close colleagues, especially countryman Thomas Huxley, who embraced his revolutionary theory were busy defending it against religious leaders and many scientists as well. Religious belief among scientists was common then, and the ideas in *Origin* seemed to attack the very basis of their beliefs—as of course it did. Even Alfred Wallace could not bring himself to fully accept the implications of evolution via natural selection when it came to humans. Darwin had tried to avoid offending his believer friends, and especially his wife, Emma, by skirting the issue of human evolution and implications for human nature and religious morality in *Origin.* However, he knew he would have to confront his believer critics on this most delicate issue, and after a decade of defending his theory of evolution following the publication of *Origin,* Darwin decided to attack the issue directly by writing *Descent of Man.*

More than three decades after he developed the key idea to understanding evolution, Darwin decided he had to deal directly with the issue of human evolution. His *Descent of Man* would finally confront the implications of his theory for humans. In *Descent* he detailed the evidence of man's descent from "lower" forms of life and compared the mental and moral traits of humans to other animals. His conclusions, unobstructed by religious doctrine, are plainly set forth in *Descent* and show clearly that humans, like other species, are descended from preexisting life forms. Darwin's own words from the "General Summary and Conclusions" in *Descent of Man* are used here to convey his sense of purpose in the work:

> The main conclusion here arrived at, and now held by many naturalists who are well competent to form a sound judgment is that man is descended from some less highly organized form. The grounds upon which this conclusion rests will never be shaken, for the close similarity between man and the lower animals in embryonic development, as well as in innumerable points of structure and constitution, both of high and of the most trifling importance—the rudiments which he retains, and the abnormal reversions to which he is occasionally liable—are facts which cannot be disputed. They have long been known, but until recently they told us nothing with respect to the origin of man. Now when viewed by the light of our knowledge of the whole organic world, their meaning is unmistakable. The great principle of evolution stands up clear and firm, when these groups or facts are considered in connection with others, such as the mutual affinities of the members of the same group, their geological distribution in past and present times, and their geological succession. It is incredible that all these facts should speak falsely. He who is not content to look, like a savage, at the phenomena of nature as disconnected, cannot any longer believe that man is the work of a separate act of creation.

Showing that humans are closely connected to and share common ancestors with what Darwin called the lower animals was the goal of *Descent of*

Man, and anyone who reads the book with a reasonably open mind will agree that Darwin makes a convincing case. Of course closed-minded believers then and now are unmoved by evidence and logic and choose to interpret religious stories in narrow, literal ways.

Within a few days of the publication of *Descent* a strong critique of natural selection was published and it proved to be popular among believers who wanted to convince themselves that Darwin and his scientific ideas were wrong. George Mivart's *On the Genesis of Species* (1859) appealed to grand-design arguments, made light of half-evolved wings that could not fly, pounced on differences among Darwin and his close colleagues, and suggested that Darwin was overstepping the proper domain of science into religion by connecting man to other animals. Mivart's critique and others that followed accused Darwin of undermining civilization by connecting man so closely to other animals. What would become of morality? Rather than retreat from the storm of criticism, Darwin worked hard on another book designed to continue what *Descent* had started, and within about a year, in 1872, *The Expression of Emotions in Man and Animals* was completed and published.

In *Expression* Darwin showed the many facial expressions that are common to man and other animals and from these and other behavioral traits inferred that many feelings are similar too. The human species was now more closely linked to the rest of the living world than ever before. Neatly separating humans from the great apes and other mammals now required that one ignore mounting evidence in favor of natural selection, but many believers then and now accomplish this feat with ease. When one's religious beliefs, learned as a child and reinforced as an adult, are perceived to be threatened it is one's duty as a believer to resist the threats using any means necessary. Anything short of this might result in bad things happening in one's life on Earth, or worse, in the hereafter.

During the final decade (1872–82) of his life Darwin continued to work as a naturalist, mostly studying plants and avoiding writing more on the implications of his theory of evolution for humans, but he was confident that future scientists would answer important questions facing evolutionary theory. The age of the Earth, the nature of genetic inheritance, the source of genetic variation, and many more discoveries in the twentieth century all support his theory of evolution. Although he was, and still is, despised by many religious authorities and their flocks, Darwin is now seen as the most important biologist, and perhaps the greatest scientist, of all time.

ALBERT EINSTEIN AS REVOLUTIONARY PHYSICIST, PACIFIST, AND FREE THINKER

Three years before Darwin died, Albert Einstein was born (on March 14, 1879) in Germany to Jewish parents who did not follow the Jewish faith and customs closely. Questioning authority was to become a lifelong tradition for Einstein. Highlights of his revolutionary ideas in physics, his strong commitment to pacifist ideals, and his free-thinking spirit are presented in the following sections.

Becoming a Physicist

Albert Einstein disliked the mindless discipline practiced at the Catholic elementary school he attended as a child and the gymnasium as an adolescent. Although slow to speak as a child, when he was introduced to a geometry textbook at the age of twelve Albert quickly showed his talent for mathematics. Within a year he was reading other mathematics books and soon decided that the physical sciences were far preferable to biology because ideas could be expressed mathematically rather than in ordinary language. At about the same time that he was becoming immersed in mathematics he was questioning all sorts of authority, including that of religion and government. Mathematical order and logic were far preferable to revealed religion and governmental laws in learning the truth of the world.

Einstein left the gymnasium in Munich before he could graduate, and at the age of sixteen, in 1895, Albert enrolled in the Swiss Federal Polytechnic School (Eidgenossische Technische Hochschule, or ETH) in Zurich. However, difficulties with the entrance exam forced him to study for a year at a nearby school; he was finally admitted to ETH in 1896. During his four years at ETH he concentrated on mathematics and physics and in 1901 became a citizen of Switzerland. Soon afterward he started work in a Swiss patent office and continued his study of theoretical physics. Einstein worked on a number of theoretical projects, including a method of determining molecular sizes, and he submitted his work as a dissertation to the University of Zurich in July 1905. It was accepted two months later, and for Einstein 1905 would become the most important year of his life. Not only did he receive a doctorate in physics from the prestigious University of Zurich, he produced five remarkable papers that would soon transform the field.

Einstein's Big Year

In terms of publications, while Charles Darwin's big year was 1859 Einstein's was 1905. The mechanism of natural selection put forth by Darwin to explain the origin and great diversity of species unified biology as never before. Einstein's work, published in 1905, similarly unified and extended disparate phenomena in physics and quickly brought him to the attention of theoretical physicists. Two of the papers dealt with molecular motion, two developed the foundation of special relativity, and the last involved the energetic properties of light in what we now call quantum theory. It was this last paper and subsequent work related to it that are closely tied to Einstein's winning the Nobel Prize for physics in 1922. And like Isaac Newton's year of 1666, when he unified and extended classical mechanics and developed related mathematical methods that we now call calculus, Einstein's year of 1905 is equally remarkable. In each case physical laws governing nature were greatly unified and new conceptions of the physical world were developed.

The impact on society of Darwin's *Origin* was immediate and huge. Einstein's 1905 papers had considerable impact within the small world of theoretical physics, but it was some time before the public began to learn of relativity and quantum theory, and even then they seemed to have little practical significance for the average person. The implications of Darwinian evolutionary theory for our understanding of human nature and our relationship to primates and other animals were quickly recognized by those who understood *Origin*. The mathematical, abstract nature of Einstein's work limited direct access to his papers to physicists and mathematicians. Eventually, his ideas became better known to the public, and his famous equation ($E = mc^2$) relating mass to energy is widely used in the media to portray the technical side of science. It is also associated with the building of the atomic bomb during the Second World War.

On Religion and Related Things

Einstein's ideas on religion are discussed in a variety of sources, but one of the best is *Einstein and Religion* (1999), by physicist Max Jammer. Albert's parents did not practice Jewish rites, and this undoubtedly allowed him to question religious authority far more easily than would have been the case if he had been raised to follow religious dogma. He knew of both Jewish and Catholic rituals and stories but was more influenced by his love of nature and music than by traditional religious dogma. Einstein developed what he called a *deep religiosity*, but it was not associated with traditional religious training. His spirituality or religiosity was more closely associated with the wonder and beauty of the workings of nature. He did not believe in the personal God that is at the core of most religious teaching, but saw God in strictly deterministic, nonanthropomorphic ways. A statement of Einstein's describes what he called his cosmic religious feeling:

> The individual feels the futility of human desires and aims and the sublimity and marvelous order which reveal themselves both in nature and in the world of thought. Individual existence impresses him as a sort of prison, and he wants to experience the universe as a single significant whole. The beginnings of cosmic religious feeling already appear at an early stage of development, for example, in many Psalms of David and in some of the Prophets. Buddhism, as we have learned especially from the wonderful

writings of Shopenhauer, contains a much stronger element of this. The religious geniuses of all ages have been distinguished by this kind of religious feeling, which knows no dogma and no God conceived in man's image; so that there can be no church whose central teachings are based on it. Hence, it is precisely among the heretics of every age that we find men who were filled with this highest kind of religious feeling and were in many cases regarded by their contemporaries as atheists, sometimes also as saints. Looked at in this light, men like Democritus, Francis of Assisi, and Spinoza are closely akin to one another. (Quoted in Jammer, 1999, pp. 78–79).

By rejecting the more common religious ideas based on fear and power, in which gods are viewed in anthropomorphic or humanlike terms, Einstein's cosmic religious feeling was seen as a driving force behind his scientific research. To the religious establishment, Einstein's rejection of a personal God in favor of a cosmic religious feeling that he related to nature's mysteries or laws, was unacceptable, and he was widely criticized when he made his ideas public in an essay in the *New York Times Magazine* on November 30, 1930. However, this concept of religion avoids the fundamental conflicts between science and religion that are viewed as inevitable when a personal God who intervenes at will is at the center of religious beliefs. Einstein called the common conception of a personal God:

Not only unworthy but also fatal. For a doctrine which is able to maintain itself not in clear light but only in the dark, will of necessity lose its effect on mankind, with incalculable harm to human progress. In their struggle for the ethical good, teachers of religion must have the stature to give up that source of fear and hope which in the past placed such vast power in the hands of priests. The further the spiritual evolution of mankind advances, the more certain it seems to me that the path to genuine religiosity does not lie through the fear of life, and the fear of death, and blind faith, but through striving after rational knowledge. In this sense I believe that the priest must become a teacher if he wishes to do justice to his lofty educational mission. (Quoted in Jammer, 1999, p. 95).

With regard to the personal God concept, Einstein can be considered an atheist or agnostic, but in the light of his cosmic religious feeling, he can be called a very religious person. Although Einstein considered himself a spiritual or religious person and rejected the label of atheist, it is easy to see why many regard him otherwise because of his criticism of the dogma of personal-God beliefs. In a June 1948 essay in the *Christian Unitarian Register* Einstein notes the strong religious conviction of great scientists:

While it is true that scientific results are entirely independent from religious or moral considerations, those individuals to whom we owe the great creative achievements of science were all of them imbued with the truly religious conviction that this universe of ours is something perfect and susceptible to the striving for knowledge. If this conviction had not been a strongly emotional one and if those searching for knowledge had not been inspired by Spinoza's "Amor Dei Intellectualis," they would hardly have been capable of that untiring devotion which alone enables man to attain his greatest achievements. (Quoted in Jammer, 1999, p. 117).

Finally in 1954, a year before his death, Einstein stated his ideas on religion very clearly in an interview with Professor William Hermanns of Stanford University:

> About God, I cannot accept any concept based on the authority of the Church. As long as I can remember, I have resented mass indoctrination. I do not believe in the fear of life, in the fear of death, in blind faith. I cannot prove to you that there is no personal God, but if I were to speak of him, I would be a liar. I do not believe in the God of theology who rewards good and punishes evil. My God created laws that take care of that. His universe is not ruled by wishful thinking, but by immutable laws. (Quoted in Jammer, 1999, pp. 122–23)

Confusion over whether Einstein was or was not religious or believed in God derives from the vagueness of the terms *religious* and *God*. It should be clear at this point that what Einstein meant when he referred to himself as religious or to his belief in God in no way involved the personal-God concepts common to most organized religions. When he said he wanted to know God's thoughts, Einstein meant he wanted to know the fundamental laws of nature. *Wanting to know God's thoughts* was merely a rhetorical device that, unfortunately, was and continues to be misinterpreted. It is important to understand Einstein's metaphorical use of religious terms when trying to understand his ideas on the relationship of science to religion, education, and other human endeavors.

Misusing Relativity and Quantum Theory

Relativity and quantum theory eventually flowed into the public consciousness but were often misinterpreted to mean, respectively, that one idea is as good as another and that accurate measurement is no longer possible. To many religionists it offered proof that science is not superior to religion in determining what is true or real. Although Einstein said repeatedly that beyond their relevant domains within physics relativity and quantum theory have no real implications for religion or philosophy or other aspects of life, many opportunists have tried to show how these new ideas could be applied to everyday life. Materialism was pronounced dead by those opposed to science, and relativism became fashionable in many areas, including anthropology, literature, and the visual arts. Mystics and believers in pseudoscience, for example astrology, creation science, phrenology, embraced the relativism and uncertainty they saw in Einstein's work to promote their own agendas designed to take advantage of people's tendencies to accept ideas without proper skepticism and inquiry. The postmodernists of the second half of the twentieth century often used the confusion surrounding relativity and quantum theory to justify relativism, the doctrine that proclaims that one idea is no better than another. This allowed creationists and postmodernists to challenge the legitimacy of evolutionary science and science in general. Of course it is difficult to judge the impact of Einstein's ideas outside the domain of science, but most observers agree that much of the impact has been the result of misinterpretation and misapplication of relativity and quantum theory. Observers within science agree that Einstein's status as the most important scientist of the twentieth century is well deserved.

BERTRAND RUSSELL AS PHILOSOPHER, MATHEMATICIAN, AND SOCIAL CRITIC

Outspoken, brash, confrontational, and *sharply critical* are terms that could be used to describe Bertrand Russell as an adult. He is without question one of the most important philosophers and social critics of the twentieth century. Bertrand's parents both died within three years of his birth (on May 18, 1872), so he and his older brother Frank went to live with their well-to-do grandparents. Bertrand was taught at home by private tutors while Frank was sent away to school, so his existence was somewhat lonely, and strictly regulated by his grandparents. After his grandfather, Lord John Russell, died in 1878, it was his grandmother, Lady Frances Russell, who oversaw much of Bertrand's daily life. She instilled in him an independent spirit that grew stronger as he reached adulthood.

Russell's study of religion during his early teenage years led him to reject belief in God and related dogma; shortly after his sixteenth birthday he confided in his diary:

> I should like to believe my people's religion, which was just what I could wish, but alas, it is impossible. I have really no religion, for my God, being a spirit shown merely by reason to exist, his properties utterly unknown, is of no help to my life. I have not the parson's comfortable doctrine, that every good action has its reward, and every sin is forgiven. My religion is this: to do every duty, and expect no reward for it either here or hereafter. (In Seckel, 1986, p. 18).

When he was eighteen he entered Cambridge University's Trinity College on a scholarship in mathematics, and after successful study in mathematics and philosophy Russell became a fellow at Trinity. This association with Trinity College would continue throughout his long life, although there was a time around the First World War when he lost his lectureship for a while because of his pacifist activities, being imprisoned for six months in 1918. Russell married Alys Pearsall Smith in 1894, and the two of them visited Germany and studied the social democrats; Bertrand published his first book, *German Social Democracy,* in 1896. With a comfortable inheritance to live on, Bertrand was able to travel with Alys, and America was prominent on his lecture tours. In 1897 he published *An Essay on the Foundations of Geometry,* and from 1900 to 1912 he published four more books, on logic, mathematics, and philosophy. His major work on logic and mathematics, *Principia Mathematica*, was published, with Alfred Whitehead as coauthor, in three volumes from 1910 to 1913.

After *Principia* Russell published about one book each year during the remainder of his life on a variety of topics, from social justice to mysticism to analysis of the mind, to Einstein's relativity, to education, to religion. No topic seemed to be outside the scope of his considerable intellect and wit, and for the many contributions he made he was awarded the Nobel Prize for literature in 1950. During the last two decades of his life Russell actively opposed nuclear weapons, and in 1955, just before Einstein's death, the famous Einstein-Russell Declaration was developed and led to important efforts to avert nuclear disaster. He was chosen as first president of the Campaign for Nuclear Disarmament, and in 1961 he was imprisoned for civil disobedience related to his CND activities. At the age of eighty-nine Bertrand and his fourth wife, Edith, were sentenced to two months in prison and served seven days before being released on medical grounds. In 1963 he formed the Bertrand Russell Peace Foundation and for the remainder of his life Russell worked for peace and social justice. His three-volume *Autobiography* (1967, 1968, 1969) was completed shortly before his death, on February 2, 1970, in Wales.

Why Russell Was Not a Christian

Long before he published *Why I Am Not a Christian and Other Essays on Religion and Related Subjects* in 1957, Bertrand Russell made very clear his disdain for religious dogma. As a teenager he saw many shortcomings of religious teachings and argued throughout his life that rational behavior offered the best path for humankind. The logical rigor that he brought to his analysis of mathematics carried over into much of his analyses of humans and their institutions, including religion. At one point, when young Bertrand was about thirteen, a tutor told him that if he believed Darwin's theory of evolution to be true, he could not also be a Christian. Although he did not see the incompatibility at the time, he decided that if he had to make a choice, he would choose evidence and logic (Darwinism) over Christianity's stories.

Until the age of eighteen Russell could not refute the *first cause* argument used to "prove" the existence of God. But then he read in John Stuart Mill's *Autobiography* that the question *Who made God?* takes care of that problem quite nicely. After that, for Russell there were no serious logical problems associated with refuting the existence of God, although his wider reading at Cambridge offered new arguments for God that he eventually worked through. Finally he ceased thinking that such arguments were worth examining.

Long after he had logically disposed of God and traditional religious dogma, Russell published what is probably his best-known work on religion, *Why I Am Not a Christian*. He showed that the term *Christian* has had many different meanings, depending on one's church affiliation, family, community, country, and when in history one has lived. He observed that Hell, for example, used to be an important part of all Christians' beliefs but in modern times one can be considered a Christian without professing a belief in eternal damnation. An argument offered by some supporters of Christianity and other religions is known as the *argument from design*. The design argument has been around since long before Darwin and goes something like this: because nature includes many complex objects, like eyeballs, wings, brains, and so on, it follows that these wonderfully complex objects must have been designed by a super-intelligent being; that is, God. Of course this is no argument at all, but it appeals to the person who lacks logical ability or is simply overwhelmed by the biochemistry or physics and so on behind these wonderful objects. In its current form the argument from design is known as *irreducible complexity*, and the object is the living cell. The amazing biochemical activity that occurs within the living cell, so the argument goes, is simply too complex to have arisen by means of natural selection. Appealing to ignorance of biochemistry, a common feature of most people, this kind of "argument" can also be called the *incredulity* argument. It seems incredible to the person ignorant of biochemistry that such complexity as that which goes on within a living cell can be explained in terms of evolutionary biology.

Another argument for God critiqued by Russell is the moral argument. Without an all-knowing God people could not know right from wrong. For Christians this means that without the Bible's moral messages people would be unable to know right from wrong. Of course how people determined right from wrong before Christianity's God appeared is not made clear by those who favor the moral argument, and careful reading of the Bible yields many inconsistencies about how one should behave in order to be a good Christian. Should one turn one's cheek when smitten, or smite back? Is it reasonable to think that a just and caring God would condemn a person to everlasting damnation simply because he chose to ignore Christ's teachings? Russell observes:

> You will find in the Gospels Christ said: "Ye serpents, ye generation of vipers, how can ye escape the damnation of hell." That was said to people who did not like his preaching. It is not really to my mind quite the best tone, and there are a great many of these things about hell. There is, of course, the familiar text about the sin against the Holy Ghost: "Whosoever speaketh against the Holy Ghost it shall not be forgiven him neither in this world nor in the world to come." That text has caused an unspeakable amount of misery in the world, for all sorts of people have imagined that they committed the sin against the Holy Ghost, and thought that it would not be forgiven them either in this world or in the world to come. I really do not think that a person with a proper degree of kindliness in his nature would have put fears and terrors of that sort into the world. (Quoted in Seckel, 1986, p. 67).

In looking at the history of Christian religion Russell observes that great faith has usually resulted in great cruelty:

> That is the idea—that we should all be wicked if we did not hold to the Christian religion. It seems to me that the people who have held to it have been for the most part extremely wicked. You find this curious fact, that the more intense has been the religion of any period and the more profound has been the dogmatic belief, the greater has been the cruelty and the worse has been the state of affairs. In the so-called ages of faith, when men really did believe the Christian religion in all its completeness, there was the Inquisition, with its tortures; there were millions of unfortunate women burnt as witches; and there was every kind of cruelty practiced upon all sorts of people in the name of religion. (Quoted in Seckel, 1986, p. 69).

In summing up his ideas about the foundation for religion, Russell identified fear as the main source:

> Religion is based, I think, primarily and mainly upon fear. It is partly the terror of the unknown, and partly, as I have said, the wish to feel that you have a kind of elderly brother who will stand by you in all your troubles and disputes. Fear is the basis of the whole thing—fear of the mysterious, fear of defeat, fear of death. Fear is the parent of cruelty, and therefore it is no wonder if cruelty and religion have gone hand-in-hand. (Quoted in Seckel, 1986, pp. 70–71).

All in all, Russell saw religious dogma as mostly harmful to humankind. He selected Christianity to critique in detail because it was the dominant brand of religious dogma in Europe and America, his main audiences throughout his long life. The mind-set encouraged by most Christian religious dogma was seen as antithetical to the mind-set required for a life guided by reason, and the historical record of religious strife simply reinforced Russell's thinking here.

Although Russell was not a scientist in the same sense that Darwin and Einstein were scientists, he understood the complexities and subtleties of science and why it offered the best hope for future humankind. Many see Russell as the most influential philosopher of the twentieth century.

Chapter Three:
Scientific and Religious Habits of Mind

Chapters 1 and 2 in this book presented some of the key ideas on science and religion of three great intellects, Charles Darwin, Albert Einstein, and Bertrand Russell. These men were raised in different kinds of families, but each was exposed to religious dogma early in life and later rejected it in favor of reason and evidence. Of the three, Darwin was probably the strongest believer of the creation story early in life and was influenced by family and friends who were also believers of sorts. His early creationist beliefs were challenged by mounting evidence gathered during his five-year voyage on the *Beagle* and by his reading of certain scientific books, especially Lyell's *Principles of Geology*, in which evidence for a very old Earth was presented. By the time Darwin realized that natural selection was the mechanism responsible for the origin of species and the great diversity of life on Earth, he had given up on the creation stories that had been part of his thinking for about thirty years. The God hypothesis was simply not necessary to explain the dynamics of nature, so the role of God in the natural world was diminished considerably.

The scientific habits of mind developed by each of these outstanding thinkers eventually ruled out the simplistic concept of a personal God as portrayed in the Bible and other holy books. Russell's scathing criticism of Christianity, in particular, was the strongest and most direct of the three, seeing organized religion as the cause of great suffering over the years. Comparing the habits of mind associated with scientific thought and embodied by persons like Darwin, Einstein, and Russell to the habits of mind embodied by religious fundamentalists and others who believe in a personal God who intervenes in people's lives and in nature is the main subject of this part of this book. What are these habits of mind, how are they different, and how do they interact in the mind of the average person?

Scientific Habits of Mind

A habit of mind can be described as a way of thinking that characterizes one's approach to problems or questions encountered in daily life. A scientific habit of mind is not easy to define, but it can be thought of as a tendency to view the world as knowable in real, natural terms. A scientific outlook combines skepticism toward claims of supernatural intervention with a confidence that nature, including humans, is knowable and is governed by the same basic laws everywhere in the universe. A scientific habit of mind must see knowledge about nature as open to change, and the history of science supports this claim. As new conjectures are made, evidence must be offered that can be checked carefully and objectively by others using methods agreed upon by the scientific community as valid and reliable. Eminent evolutionary biologist Ernst Mayr puts it this way: "One of the most characteristic features of science is this openness to challenge. The willingness to abandon a currently accepted belief when a new, better one is proposed is an important demarcation between science and religious dogma." (NAS, 1998, p. 43). This openness in science is characterized by making new conjectures and refuting earlier ones, resulting in a process of trying to make a better fit between new evidence and existing theory. Mayr goes on to liken this process to evolution of life via natural selection: "Indeed it is by a Darwinian process of variation and selection in the formation of hypotheses that science advances." (p. 43).

Keeping one's mind open to new evidence is clearly an important habit of mind that is central to scientific progress in better understanding nature's laws. Sometimes the openness characteristic of science is described as a "tentativeness" of scientific knowledge. All scientific knowledge is open to change; physicist Roger Newton compares science with religion in this regard:

> The opponents of evolution often contend that it is, after all, "only a theory." While this is correct, and from the point of view of scientists quite innocuous—even the motion of the earth around the sun is, in a sense, "only a theory"—it carries a profound meaning for religious fundamentalists, for whom the account of the Bible has the authority and certainty of scripture, which science can never provide. Here is precisely the point of fissure between science and religion. (1997, p. 79).

Maintaining an open mind in science means that all theories are considered tentative or open to change, even though many scientific theories (for example, atomic-molecular theory of matter, evolution of life via natural selection, plate tectonic theory) are considered to be very firmly established by overwhelming evidence.

Another aspect of the scientific habit of mind is informed skepticism. Russell is one of the great skeptics of the twentieth century, encouraging rational doubt and a spirit of critical receptiveness to new ideas. Maintaining a balance between skepticism toward new and different ideas and openness to these same ideas requires a careful, rational approach. Strong biases must be temporarily set aside in order to fairly examine knowledge claims that may seem incredible at first. After all, much of current scientific theory seemed incredible when first introduced by people like Darwin and Einstein. This balance between skepticism and openness in science is described by the important science education document *Science for All Americans* in this way:

> Although a new theory may receive serious attention, it rarely gains widespread acceptance in science until its advocates can show that it is borne out by the evidence, is logically consistent with other principles that are not in question, explains more than its rival theories, and has the potential to lead to new knowledge. Because most scientists are skeptical about all new theories, such acceptance is usually a process of verification and refutation that can take years or even decades to run its course. (AAAS, 1989, p. 186).

Questioning or mistrusting arguments from authority is another aspect of the scientific mind-set, and because most people want certainty in their lives this habit of mind is especially difficult to achieve. Until Copernicus and Galileo dared to question the authority of the Catholic Church, it was decreed that Earth was the center of all things. As God's most important creation, the thinking went, Man would quite naturally be placed at the center of all things. To question this dogma was to question the authority of the religious leaders, and blasphemous questioners were put in their place and sometimes given the ultimate penalty of death. When Darwin came along with his theory of natural selection as the cause behind the great diversity and origin of species, it seemed to many religious leaders and their flocks that God was being shoved out of the picture. Requiring that those in positions of authority prove their contentions like everybody else is an attitude that can cause problems for the questioner, whether those in power are religious leaders, politicians, CEOs, or project directors. The free thinker or skeptic places evidence over authority regardless of the source of the authority. Open inquiry requires such an attitude, or habit of mind, so there are no forbidden questions.

Finally, curiosity and a strong need to know are required to understand nature's secrets more accurately and deeply. The drive and determination to stay with a problem until a satisfactory solution is attained often sorts out the more successful from the less successful scientists. Persevering in the face of difficulty is a characteristic that is not unique to scientific inquiry, but it is usually associated with important scientific discoveries simply because seeing things from a new perspective is not easy for humans to achieve. It really does appear to Earth-bound observers that all objects in the sky move around us, and developing a new perspective on things is not easy.

British biologist Lewis Wolpert (1992) describes scientific thought as *unnatural.* This unnatural, uncommon sense required of scientists and others who understand scientific theories deeply is very different from the common sense or folk knowledge that seems to develop effortlessly within a community. Our sensory apparatuses and related cognitive abilities evolved as they did to optimize survival in certain physical environments, not to allow for the invention of abstract theories about the subtle laws of nature. The fact that science took so long in human history to develop and take hold supports the thesis that the scientific outlook is unnatural compared to most human activity, such as rearing children, hunting and raising food, creating shelter, and creating supernatural spirits to explain the mysterious and to offer the promise of hope in hopeless situations. Becoming accustomed to the unnatural nature of science, often at odds with our commonsense beliefs about our natural environment, is required, along with the habits of mind identified earlier, in order to feel more comfortable with science. For science educators familiar with the large database of research on *science misconceptions* it comes as no surprise to hear that science is not an easy, natural human activity. For most children and their parents, the scientific outlook is seldom developed accurately or in depth. Misconceptions about scientific theories of nature and about the nature of science itself are widespread and deep, and the religious mind-set, described in the next section, is partly to blame.

RELIGIOUS HABITS OF MIND

Unlike scientific habits of mind, religious belief seems to be quite natural. People all over the world hold various religious beliefs and have done so for thousands of years. Belief in the supernatural, which includes most religious beliefs, allows one to believe something without the need for evidence, at least not the kind of evidence required in the natural sciences. Believing that unseen gods (seldom goddesses since Christianity appeared) are behind the visible universe acting as causal agents seems to be a very easy, natural thing to do. Whether religious belief is invoked to reduce life's anxieties or to enhance one's position of power, it seems to occur naturally in nearly all societies.

Trying to explain why people believe in the unbelievable and why certain religious beliefs survive and thrive in an age of science, anthropologist Pascal Boyer (2001, p. 300) observes that "most religions routinely flout the requirement of consistency." Humans, he explains, have a "whole menagerie of mental processes that apparently conspire to lead us away from clear and supported beliefs." Derived from many psychological studies, these self-delusional processes include *false consensus effect, memory illusions, confirmation bias*, and *cognitive dissonance reduction*. These and many more departures from scientific thought help to explain why belief in the unbelievable is so common and natural. Of course the proximal cause of specific religious beliefs is early indoctrination of young children by religious adults, older siblings, and peers. The pressure to conform in order to be loved and accepted within a family or other group is very strong, and religious leaders recognize and take advantage of this fact.

Although the main purpose of this section is to describe common religious habits of mind, it is difficult not to wonder about the cognitive origins of religion and related habits of mind. Michael Persinger was one of the first neuroscientists to try to explain *God beliefs* and *God experiences* in terms of brain function, saying that "the behavior of the temporal lobe epileptic has been characterized by the persistent theme of religiosity. Like the committed preacher or the proselytizing prophet, they have a sense of the special—their experiences are somehow exceptional" (1987, pp. 19–20).

Personal revelations, whether experienced as part of a religious event or as a temporal lobe seizure, seem completely real to the brain's owner. Neuroscientists like Persinger and social scientists like Boyer see religious habits of mind to be quite compatible with brain structure and function. Those with strong God beliefs are convinced that they have had God experiences, and they participate in regular rituals such as church attendance, prayer, and missionary work to maintain and strengthen their God beliefs. These religious behaviors can give believers hope in a better life now and after their life on Earth is over.

The habits of mind associated with most religious beliefs include faith in the authority of holy books and religious leaders. Accepting without evidence what a holy book or a religious leader says is one of the hallmarks of the religious habit of mind. To challenge the basic assumptions of one's religion is often the beginning of the end of those particular beliefs, as we saw with Darwin as he gradually changed from creationist-believer to evolutionist-unbeliever during a period of about seven years.

When Galileo questioned the accepted religious dogma of a geocentric universe he was accused of heresy by religious officials and sentenced to house arrest for the remainder of his life. When Darwin argued that humans are descended from other primates, and ultimately from the simplest of organisms, he was ridiculed by church officials and other followers of church doctrine for his heretical claims. And when John Scopes taught about human evolution in his high school class in Dayton, Tennessee, he was tried and convicted in 1925 of violating a state law that outlawed the teaching of human evolution in public schools. As a habit of mind, questioning religious authority has had severe consequences throughout recorded history. At the very least one is criticized by religious authorities and their followers.

To say to the religious person that he or she has no evidence to substantiate claims of a God doing miraculous things may do little to influence the true believer who has the evidence of God experiences. Searching for causes of events in the environment is a natural trait of *Homo sapiens*, probably because it increases the likelihood of survival of their genes. This interest in causality seems to be central to both scientific and religious habits of mind. However, for most religions it is the imagined supernatural world rather than the natural world that contains the causal agents. Belief in a personal God who intervenes and causes things to happen in the real/natural world is the most common conception of God, at least in the United States and other countries where prayer is a religious ritual. Belief in a God who created the universe but does not intervene into the natural world poses far fewer problems for the scientist who must assume that the natural world operates according to natural laws.

The habits of mind associated with religion vary because of the wide variation in religious beliefs, but on the whole, if the belief system involves a personal God who intervenes in people's lives or in the natural world in general, conflicts with science will occur. The religious mind-set of unquestioning acceptance of predetermined doctrine must inevitably conflict with a scientific mind-set that requires open inquiry into all parts of the natural world, including people's minds.

POINTS OF CONFLICT

It is sometimes argued that no real conflict needs to exist between science and religion. Science deals with the natural world and religion with the supernatural, and because these domains do not overlap, so the argument goes, there is really no conflict. One of the problems with this argument is the vague definition of religion. Just as God concepts vary widely now as they have in the past, so do religious doctrines. Some believers fear Hell and eternal damnation while others do not believe in the concept of Hell. Some believe that homosexuality is a sin while others do not. Some believe that women's proper role is to serve men while others do not, and on and on and on. Each of the many thousands of religious groups has its own particular beliefs about what their God does or does not allow and what pleases Him (seldom Her). Because almost all of these believers are sure that there is only one God it must be very confusing for Him to know what to do in light of all this. So we see that the words *God* and *religion* have as many different meanings as there are believers. This should not be too surprising, since the supernatural world, by definition, is beyond inspection by mere mortals on Earth. Any person can make a claim about their God or their religion without fear that it will be discovered to be untrue, unless of course a holy book or religious leader declares it so. Unless a religion declares that Earth is the center of all things or that Earth is only ten thousand years old or that humans are not closely related to other primates or in some other way enters into the natural world of science, believers can construct supernatural worlds to their hearts' content without fear that they will be proven wrong. God's personal appearance, His likes and dislikes, the temperature of Hell, the nature of Heaven, and so on can be argued by believers until Hell freezes over and no person can provide evidence to the contrary.

Conflict between science and religion can occur even when religion makes no claims about the nature of the real world. We saw earlier that religious habits of mind are generally very different from scientific habits of mind. The open-inquiry approach of science is very different from the appeal to holy books and religious authority that believers are trained to accept at an early age. Scientific theory is ultimately based on evidence gathered in the natural world using methods that can be replicated by others. If a young-Earth creationist says, "I do not believe the Earth is billions of years old," it means that that person rejects the scientific evidence offered by scientists, even though no counter-evidence can be offered other than a statement from a holy book or a religious authority. For the young-Earth creationist in this example a point of conflict clearly exists even though the domain of the supernatural can be shown logically to be separate from that of the natural world. It is in the mind of the young-Earth creationist, not in set theory or logical analysis, that the real conflict exists. The religious person's attitude toward evidence was probably influenced by early religious training; whenever something seems to conflict with that training, the source of conflict is rejected. In some cases a religious *conversion* in adolescence or adulthood can explain the closed habit of mind represented by the young-Earth believer, but for the most part this religious habit of mind is developed during childhood before a person is capable of rational thought. The point is that the student cannot simply separate religious beliefs or habits of mind from scientific evidence. They naturally intersect when it seems to the person that they conflict. It is what happens in the person's brain/mind that is of real interest here, and for that reason a brief look at how the mind/brain works is in order.

HOW THE MIND WORKS

The title of this section also happens to be the title of an influential book by evolutionary psychologist Steven Pinker. *How the Mind Works* (1997) is a synthesis and interpretation of research into our mental life using perspectives gained from evolutionary biology, cognitive psychology, artificial intelligence, and neuroscience. The conception of mind presented by Pinker in *How the Mind Works*, by philosopher Daniel Dennett in *Darwin's Dangerous Idea* (1995), and by others who take science (evolutionary biology in particular) seriously is very different from dominant mind theories of the past, such as behaviorism and Freudianism. The human mind is seen as synonymous with brain function, and the brain is viewed as a biochemical processor of symbols. Pinker describes the mind in a section he calls "Reverse-Engineering the Psyche":

> The mind is a system of organs of computation, designed by natural selection to solve the kinds of problems our ancestors faced in their foraging way of life, in particular, understanding and outmaneuvering objects, animals, plants, and other people. The summary can be unpacked into several claims. The mind is what the brain does; specifically, the brain processes information, and thinking is a kind of computation. The mind is organized into modules or mental organs, each with a specialized design that makes it an expert in one arena of interaction with the world. The modules' basic logic is specified by our genetic program. Their operation was shaped by natural selection to solve the problems of the hunting and gathering life led by our ancestors in most of our evolutionary history. The various problems for our ancestors were subtasks of one big problem for their genes, maximizing the number of copies that made it into the next generation. (1997, p. 21).

From this description one can see how different Pinker's conception of the human mind is from previous dominant explanations. Beliefs and desires in the computational theory of mind are configurations of symbols that are processed by various modules of the brain. Pinker points out that the computational theory of the mind is different from the computer-as-mind metaphor and that without computational theory the evolution of the mind/brain is not possible.

Although brain structure is initially determined genetically in the newborn baby, the environment after birth is very important in influencing continued development of the various modules and functioning of the brain. The interaction of the brain, through sensory input, with the environment goes through critical periods when brain development can be hampered if proper sensory input is not received; these critical periods are well known for humans and other animals.

Because religious belief is tied so closely to emotion, it is important to understand the role of emotions in the mind/brain if we hope to better understand the relationship between religious and scientific habits of mind. Pinker suggests that the main role of emotions is to help select a single goal for action from among many that face us at any given time:

> The emotions are mechanisms that set the brain's highest-level goals. Once triggered by a propitious moment, an emotion triggers the cascade of subgoals and sub-subgoals that we call thinking and acting. Because the goals and means are woven into a multiply nested control structure of subgoals within subgoals within subgoals, no sharp line divides thinking from feeling, nor does thinking inevitably precede feeling or vice versa. . . . Most artificial intelligence researchers believe that freely behaving robots . . . will have to be programmed with something like emotions for them to know at every moment what to do next. (1997, pp. 373–74).

So emotions are an important part of the cognitive machinery that has evolved to keep us alive in primitive hunting and gathering environments. Emotions like fear, disgust, surprise, and anger are associated with certain universal facial expressions and actions in response to certain environmental conditions. The function of each emotion is to prepare the mind to seek actions that optimize the chance for survival. The more general emotion of unhappiness is a signal to search for changes that will lead to happiness, the overall condition that optimizes survival over the long haul.

The most talked-about emotion, love, is sometimes given another name by biologists: altruism. Helping another person at expense to you is at the center of love and is what binds family and friends together. In most religions it is love that is used as the reason for helping others who may be less fortunate than you and it is love or altruism that is at the core of the so-called Golden Rule. The feeling of love is described by people who have God experiences, and the terms *God* and *love* are often used interchangeably by believers.

It was mentioned earlier that Russell identified the emotion of fear as the original source of religion. This conclusion seems logical enough: there were many things to fear in the primitive world of humans and their immediate ancestors, and something that relieved the unhappy state of fear would be welcomed by the fearful. Overcoming a state of fear through certain rituals could be a big advantage in a dangerous environment. However, survival of a family unit or a community of families might also be enhanced by altruistic behavior in which the feeling of love is used to help achieve a more cooperative community, giving it an edge over less cooperative communities. Discovering effective means to achieve this kind of cooperation could be a real source of power for some within a community. The shaman within a group wields considerable power over others who are unable to commune with the gods and fend off evil spirits.

Appealing to imaginary, supernatural beings as causal agents is especially attractive to those in need of assistance when times are tough or frightening—and times can be tough for all of us throughout our lives. Achieving a more peaceful, calm state through a belief in an unseen force that protects you might provide an edge for survival just as meditation or physical exertion or ingesting a sedative can calm one's nerves and thus provide one with an edge for survival. It is not difficult to see that the meme (Dawkins, 1976) of religion could be an advantage in the struggle to survive in the face of difficult, threatening environments.

It is when this unit of culture (meme) becomes organized as a source of power utilized by a few over many that things seem to go awry. The authority of religion often becomes so dominant that even the family unit is seen as a threat to religious leaders' power, and sons and daughters are sometimes encouraged or coerced to break all ties to their family in order to join the religious family directed by the all-powerful shaman. Religious groups continually splinter and form new groups with somewhat different beliefs, and each new group is headed by a new shaman or priest who professes special powers or insights into the supernatural world of the gods. The religious mind-set requires strict adherence to rules set down by shamans or found in holy books handed down by other religious leaders.

In *Religion Explained: The Evolutionary Origins of Religious Thought* Pascal Boyer (2001) identifies a number of mental processes that are common in humans and lead us to accept the illogical, unbelievable claims of religious dogma:

- *Consensus effect* occurs when people adjust their memories and judgments to coincide with the consensus of the group.
- *Memory illusions* can be created so that people believe that implanted memories really occurred.

- *Confirmation bias* leads people to confirm what they already believe to be true. Confirming data are remembered and counted while disconfirming data are not.
- *Cognitive dissonance reduction* biases people to readjust memories of past experiences in light of new experiences, causing them to think that they had a particular impression all along.

These and many other mental biases have been found to cause the mind to work in ways that depart from rational, scientific thought. When these biases are combined with the previous interpretation of religion and its origins, it is not difficult to see why religion in its various forms is so dominant in cultures present and past. The mind/brain seems to be biased toward religious belief and away from scientific thought.

More on Points of Conflict between Religion and Science

Although logical analysis of the separation of the natural world of science from the supernatural world of religion can reveal the two worlds as separate or non-overlapping, analysis of the mind/brain shows how the two worlds can become entangled. The more emotional, feelings-oriented world of religion can become entangled with the rational world of science in the mind of a person who is exposed to both during childhood and adulthood. As conflicts arise between current beliefs and new evidence it is natural to try to settle the conflicts in one's mind by choosing what seems to be the best option at the time. However, one's memory is not divided into separate religion and science compartments, and as we saw from Boyer's examples of mind biases it is very difficult to be logical and objective even when religious doctrine is not hanging over one's head. If the concept of *angel* becomes implanted into one's mind as a child it can seem as real as the concept of *moon* even though one is a real object out there while the other is not. Once *angel* becomes part of one's memory traces it does not simply disappear as one grows older, although it may be modified in various ways according to later experience, including acquiring a scientific education.

So the supposed non-overlapping spheres of science and religion actually do overlap in the real world of memory traces in the mind/brain, and since education involves changing the complex pattern of memory traces the potential for conflict exists. Our memory of our life's experiences consists of a complex web of memory traces that have the potential to interact in ways that are often beyond our control. Even though we can logically separate the natural world of real objects out there from the supernatural world of angels and so on, our mind can combine and recombine the two worlds in complex ways. The following letter from a college student to her instructor (me) in a physics course is an example of the conflict experienced by students who try to learn scientific knowledge that seems inconsistent to religious believers who envision a literal interpretation of certain stories in holy books. The science content in question is the Big Bang theory in physics, although the student also brings Darwin's theory of evolution of life into her comments.

> Dr. Good:
> While I studied the chapter and wrote the answers according to the textbook, I have to disagree with the answers I wrote on the exam. I do not believe that some millions of years ago, a bunch of stuff blew up and from all that disorder, we got this beautiful and perfect system we know as our universe. You do not get order out of chaos and you do not get something out of nothing, so to believe that our universe and our world blew up out of particles is inconceivable to me. The structure and perfect way our solar system functions could not have come from an explosion.
> I personally believe in God as a supreme being. I cannot imagine living my life believing we are just here to live and die on some planet. I believe that God loved you and me so much that He personally created this beautiful world for us to enjoy. The world is too perfect for there not to be a God who created it.
> I do not understand how a person cannot have faith. I have to believe in something. I have never seen the wind but I have felt and seen the effects of the wind. The same is true with God. I have seen the effects of God and one of those is the universe. God is always there for me and I have felt His Love many times. You said in class that the Big Bang was the beginning and everything has to have a beginning. I believe that God created the earth and that that was the beginning. The Bible says in Genesis that in the beginning there was God. He doesn't have a beginning or an end, He has always been here and He will always be here. That is one of those things that our minds cannot comprehend; we just have to have faith and believe. He was there in the beginning and He created the beginning. I heard this joke one time. This man thought he was pretty powerful so he decided to challenge God. He told God "I can do everything you can do." God said all right, make a wind strong enough to blow down that tree and the man blew really hard and the tree fell down. God said, all right make a man, so the man bent down and started forming a man from dirt but God said, No I meant make your own dirt. The Bible says in Revelations 4:11 that God created everything and it is for His pleasure that they exist and were created. To say that the universe "just happened" or "evolved" requires more faith than to believe that God is behind the complex organization of our solar system.
> I believe 100% in the Biblical account of creation. I believe the earth, as we know it is only 6,000 years old. Scientific evidence may suggest otherwise, but God created a mature world. He created Adam the man, not a fetus. He created huge trees, not tiny seeds in the ground. He created huge canyons and mountains that to our simple minds must have taken thousands of years to form. He created beauty and perfection in 6 days that having started from scratch might have taken the assumed millions of years. He formed things with His hands in a second that nature would have taken years to form. And with those same powerful hands He is reaching out to us. We will never know all the answers to how God created the Earth, but we do know that He created because He loves us.

> I would like to thank you for presenting your beliefs and for challenging mine. I respect your position, but my faith in my beliefs has not wavered. I would like to challenge you to expand your beliefs to where they are not just based on knowledge, but a belief based on a faith in God. This class has been interesting and I will not forget what I have learned.
> *Sincerely,*

This letter represents a respectful *thanks but no thanks* approach to learning science. If it was actually written by the student rather than by a church official or another true believer, she accepts without question the authority of her Bible's interpreters. Her role is to accept rather than to question, and the letter suggests that she does this very well. Respect for evidence and the tendency to question are clearly lacking even though she seems to respect my *beliefs* about scientific ideas, inviting me in the final paragraph to expand my beliefs based on a faith in her God. The student delivered the letter to me in person at the end of the semester while another student was in my office so I was unable to talk to her about it, but it seems clear that her religious habits of mind were in conflict with whatever scientific habits of mind she possessed—and the latter lost out to the former. The default rule seems to be *When there is a conflict between religious beliefs and science, religion wins.* Adjusting religious beliefs to accommodate scientific evidence, as the Catholic church, for example, has often done since Galileo and Darwin, is not allowed in this student's religious mind-set.

CONFLICTING RELIGIOUS BELIEFS: A SHORT HISTORY OF GOD

Unlike science, which seeks rational consensus on how nature operates, religious belief seems to vary according to who speaks loudest or with the greatest compassion. The so-called monotheism of today's Christianity, Islam, and Judaism evolved from the polytheism of ancient cultures. Belief in a single God (monotheism) suggests that the believers are somehow in greater agreement about the nature of their God than are believers in many gods. However, this is not the case. Variation in God/religious beliefs is mostly a function of the nature of different cultures rather than the number of different gods in charge of things behind the scenes. Karen Armstrong's *A History of God* (1993) identifies many of the God concepts imagined by true believers over the centuries before and after Christ appeared in the Bible's stories. She suggests that the concept of God is so variable as to be beyond agreement by the many different religious sects, but the fact that the idea of God continues to exist in all cultures suggests something important about human nature. The prediction by many intellectuals of the eighteenth and nineteenth centuries that the concept of God as personal protector and savior would wither and die as science took hold has not come to pass. Especially in the United States, most people still profess a belief in a personal God and say they pray for certain favors, particularly during difficult times for them or their loved ones. After the 9/11/01 terrorist attacks in New York City and Washington, D.C., politicians and others invoked God's name more than usual in public talks, and the average citizen did the same privately. Praying to the same God who is in charge of all things, including the terrorists who guided the planes into the World Trade Center towers, millions of believers asked for relief of one sort or another from *their* God. In the name of God the terrorists killed thousands of innocent civilians while in the name of God others plotted revenge against their kind. It must have been confusing to this God to have so many different kinds of prayers coming His way.

The seriousness of this issue now, and at other times throughout history during which so many have suffered and died in the name of God, should show believers that the many different God concepts people hold depend on their unique circumstances, but this is not the case. If anything, believers' ideas are held more tightly and their minds are more closed to depersonalized God concepts. In *Why God Won't Go Away* (2001), neuroscientist Andrew Newberg and his colleagues conclude that the religious impulse is rooted deeply in the biology of the brain, supporting earlier work by Persinger (1987) and other neuroscientists interested in the persistence of God beliefs. The many varieties of religious beliefs held by people now and in the past seem to offer the believer advantages similar to nonreligious meditation, reducing feelings of anxiety and depression that can be harmful to one's health. From the standpoint of evolutionary biology, reducing the harmful effects of anxiety and depression increases the fitness of the individual, providing an edge in the struggle for survival.

So the history of God, seen from this neuroscience viewpoint, is simply a history of the development of the mind/brain, where various brain functions, if they work in helping to ensure survival, become part of the body's machinery. The countless variations in God beliefs are relatively unimportant from the standpoint of evolutionary biology. What is important is the fitness benefit gained by the reduction in anxiety and depression. Really believing that we can control our lives in the face of overwhelming difficulty by appealing to unseen forces, while perhaps logically and scientifically absurd, may be advantageous in evolutionary-biology terms. The many stories associated with various God beliefs may be simply the result of individual perceptions that become what Dawkins (1976) calls *memes*, cultural practices that increase the fitness of the individuals and groups involved. Boyer (2001) uses ideas from modern cognitive science, anthropology, neuroscience, and evolutionary biology to explain why God beliefs survive and why some but not all of those beliefs are common across nearly all religions. Like art and literature, religious practice varies in its specifics, but the reasons behind it and many of its characteristics are universal. All are indications of human nature, and that is reflected in the evolution of the human mind/brain.

CONFLICT RESOLUTION?

Open inquiry and questioning authority using the methods of science are habits of mind that are in conflict with habits of mind associated with most God beliefs, in which a personal God intervenes in people's lives and in nature. Science assumes that all of nature, including all that occurs in our minds, can be studied and explained in rational, real terms, and religious belief is as open to scientific study as other human traits. Can real inquiry be practiced by true believers when apparent conflicts (evolution versus special creation, for example) exist in their minds? Can meaningful learning occur if the learner does not believe that the scientific claims are true or even approximately accurate? If the previous letter by my former student is a fair indication, the answer is probably not, at least not in a way that will ensure the kind of scientific literacy envisioned by leading scientists and educators. This kind of literacy includes more than an ability to repeat facts of science. In addition to knowledge of scientific facts, the scientifically literate person understands also the nature of scientific inquiry and something of science's history and its relationship to society.

A satisfactory resolution between the conflicting habits of mind of science and religion is difficult to imagine unless both sides (for example, evolutionist and creationist) can agree to basic definitions and ground rules. One of the most basic of these is the nature and role of evidence in determining the validity of claims. If a person claims that Earth is ten thousand years old and another says it's 4.5 billion years old, what evidence does each use to support the claim? Rejecting the scientific claim of 4.5 billion years means that one must reject the method(s) used to arrive at the answer, and that means that one must understand the nature of the methods of science, in this case radiometric dating of materials. Also, one should be able to show that an alternative method provides more accurate data, and for the young-Earth creationist this other method is reference to a holy book or person who claims special insight into God's mind. In other words, evidence in the world of religion is not comparable to scientific evidence gathered by probing nature. Lacking agreement on the nature of evidence, scientists and creationists must ultimately conflict on claims, about Earth's age, Big Bang theory, continental drift, fossil records, human-chimp ancestors, and so on. Perhaps more than anything else, a respect for scientific evidence is what differentiates scientific and religious habits of mind.

If there is to be some resolution of this conflict it is likely to be found in educational practice. The last part of this book looks at the problems faced by educators and considers possible solutions.

CHAPTER FOUR:
DEMOCRACY AND SCIENCE EDUCATION

The most influential American philosopher-educator was John Dewey, and his most influential work, *Democracy and Education*, was published in 1916, nine years before the famous Scopes evolution-creation trial in Dayton, Tennessee. Dewey argued that:

> The function which science has to perform in the curriculum is that which it has performed for the race: emancipation from local and temporary incidents of experience, and the opening of intellectual vistas unobscured by the accidents of personal habit and predilection. (1916, p. 230).

The open inquiry of science into nature's laws was seen by Dewey as representative of the kind of education needed to support a vibrant democracy. Freedom of thought and freedom of speech, the twin pillars of a democracy, are required for scientific inquiry, and Dewey placed scientific thought and knowledge at the center of a school curriculum.

Science-curriculum reform at the end of the twentieth century in the United States is well represented in *Benchmarks for Science Literacy* (AAAS, 1993) and *National Science Education Standards* (NRC, 1996), and both include an emphasis on inquiry that Dewey stressed in *Democracy and Education*. The earlier *Science for All Americans* (AAAS, 1989) was a guide for both *Benchmarks* and *Standards* and it argues convincingly for scientific literacy for all citizens:

> Scientific habits of mind can help people in every walk of life to deal sensibly with problems that often involve evidence, quantitative considerations, logical arguments, and uncertainty; without the ability to think critically and independently, citizens are easy prey to dogmatists, flimflam artists, and purveyors of simple solutions to complex problems. (p. vi).

A citizenry that acquires scientific habits of mind is in a much better position in a democratic form of government to make more informed decisions—and a public school system devoted to this kind of liberating education is essential.

At the beginning of the preface of this book four questions are asked:

- What are scientific and religious habits of mind?
- How do people settle into these different ways of viewing the world?
- Are these mindsets basically compatible or incompatible?
- Why is it so difficult to achieve widespread science literacy in our schools and in our society in general?

Questions 1 and 2 were answered, in part at least, by looking briefly at the religious and scientific habits of mind of Charles Darwin, Albert Einstein, and Bertrand Russell. From these and other considerations question 3 was answered: religious habits of mind that encourage acceptance of authority and a belief in supernatural agents that intervene in our world are fundamentally incompatible with scientific habits of mind that encourage open inquiry of and skepticism toward claims of supernatural causes in nature. To answer the fourth question it is necessary to remember what Lewis Wolpert (1992) calls the unnatural nature of science.

IS SCIENTIFIC THOUGHT UNNATURAL?

Why did science arrive so late on the scene, only a few centuries ago, and why do so few adults achieve the level of scientific literacy recommended by *Benchmarks for Science Literacy* and *National Science Education Standards*? These two questions are closely related and the answers are closely related to our earlier discussion of scientific and religious habits of mind. The short answer to both questions is because scientific thought is difficult to achieve and maintain in the face of easier, more natural ways of behaving. Built-in cognitive biases against scientific thought, like confirmation bias and overgeneralization, are strong influences and when combined with other factors in society they can work against the development of scientific habits of mind. The focus of this book has been the nature of and relationship between scientific and religious habits of mind; that theme is continued here in Chapter 4 as we try to answer the fourth question above and look at common problems facing those countries that want to achieve higher rates of science literacy among its citizens.

Few leaders in educational reform efforts represented by *Benchmarks* and *Standards* want to acknowledge the potentially negative impact of early religious training on later levels of science literacy. Most see the difficulties in criticizing organized religion, preferring instead to talk of science and religion as non-overlapping domains that can coexist peacefully. Previous arguments in this book, including the example of the college student who rejected Big Bang theory in a physics class, suggest that early religious training can and often does have a negative impact on a later science education. However, very little solid research has been done to shed light on the nature of this impact, so most of the arguments offer no scientific data to support one claim or another. The area of evolution education offers the best opportunity to focus on the religious training–science learning question, and this focus is maintained throughout the remainder of this book.

NATURAL SELECTION VERSUS SUPERNATURAL CREATION

The evolution-creation battles in the United States have been well documented, and they continue as this book is written. Every effort is made by fundamentalist Christians and others who oppose Darwin's scientific theory of evolution by natural selection, and other scientific knowledge that is deemed objectionable, to keep God and His helpers in charge of things. However, each law or school board policy that is created by those sympathetic to God's rules or laws is eventually determined by U.S. courts to be unconstitutional, violating the First Amendment, which states:

> Congress shall make no law respecting an establishment of religion, or prohibiting the free exercise thereof; or abridging the freedom of speech, or of the press; or the right of the people peaceably to assemble, and to petition the Government for a redress of grievances.

The Constitution's authors saw the dangers in allowing religious beliefs to mix with government policy, using the opening ten-word phrase *Congress shall make no law respecting an establishment of religion* to prohibit such mixing. The original Bill of Rights, offered to the House of Representatives in 1779, started with this first amendment because it was seen as crucial to the success of this new, democratic form of government. Without free speech, the rest would not mean much.

Ironically, one of the arguments often used by creationists is supposedly based on the First Amendment. They argue that by including only Darwinian evolutionary theory in the biology curriculum, their "scientific creationism" is unfairly excluded, violating the First Amendment. Of course the courts see the fallaciousness of that argument and have ruled consistently that "scientific creationism" is simply religious belief dressed up in scientific-sounding terminology. For those who do not understand the nature of science in general or evolutionary theory in particular, it may be understandable that most support the *let the students decide* argument. However, when many high school biology teachers agree that creation "theory" as set forth in the Bible should be given its fair share of time in the biology classroom, there is something seriously wrong. Opinion polls and other research show that about a third of high school biology teachers believe it is okay to teach creationism as an alternative "theory" to Darwinian evolutionary theory. How is it possible that biology teachers who have taken many college biology courses believe this? Furthermore, why is it that modern Darwinian theory, the central explanatory theory in all of biology, is given only token treatment in most high school biology classes? These questions can be answered in large part by understanding the role that religious belief plays in U.S. public education and in the larger society.

IN GOD WE TRUST AND GOD BLESS AMERICA

All U.S. currency includes the words In God We Trust, thanks to the efforts of patriotic politicians. Like Grandma and apple pie, God is used by politicians and other salespeople to market their products and ideas in this nation where about 90 percent of the citizens say that they believe in God and that religion is important in their lives. Compared with countries like England, Germany, Japan, Canada, and Norway the United States is very religious (Bishop, 1998). For example, 55 percent of the citizens in the United States say that they definitely believe in life after death, more than twice the percentage of citizens in England or Germany or Israel or Austria. About a third of U.S. citizens believe that the Bible is the actual word of God and should be taken literally, more than four times the percentage in England. In terms of claims of religious belief, the United States is clearly a very religious country. Of course it is far less clear whether St. Peter will agree when it comes time to open the Pearly Gates. What people mean when they say "I believe in God" or "God bless America," or when they repeat other religious slogans, is unclear as well, but it does seem to indicate a belief in supernatural forces of one kind or another.

In spite of the First Amendment and court decisions that are intended to keep religion out of government, we are constantly bombarded by images and words that seem to run counter to the intentions of this country's founders. All of our currency says we trust God, most of our politicians say "God bless America," our courts swear people in using the Bible or God's name, we have an official National Day of Prayer, the Pledge of Allegiance includes the phrase "under God," and on and on. These are constant reminders that religion can and does invade government. Prayers are offered before and after official government functions, so why not pray in school? Why not let religion invade public education the way religious rituals have invaded government? By allowing religion to become so entwined with government we have made it more difficult to argue that public schools should not do the same. Many religious leaders have recognized the dangers of allowing religious belief into government functions and have sided with scientists, the American Civil Liberties Union, and others who fight to keep religion out of public school science classrooms and other areas of the curriculum. However, the battle will never be over until both science and democracy are more fully understood by most citizens.

Shortly before he died of cancer, well-known scientist and science educator Carl Sagan completed a book, *The Demon-Haunted World: Science as a Candle in the Dark* (1996), that points out the dangers of scientific illiteracy to our most basic freedoms. It is an important book, in part because it relates freedom within democracy to scientific literacy. He argues:

> Pseudoscience is embraced, it might be argued, in exact proportion as real science is misunderstood—except that the language breaks down here. If you've never heard of science (to say nothing of how it works), you can hardly be aware you're embracing pseudoscience. You're simply thinking in one of the ways that humans always have. Religions are often the state-protected nurseries of pseudoscience, although there's no reason why religions have to play that role. In a way, it's an artifact from times long gone. In some countries nearly everyone believes in astrology and precognition, including government leaders. But this is not simply drummed into them by religion; it is drawn out of the enveloping culture in which everyone is comfortable with these practices, and affirming testimonials are everywhere. (p. 15).

Like John Dewey eighty years before him, Sagan emphasizes the importance of having scientifically literate citizens in a democracy. However, Dewey was not as openly critical of religious doctrine that can interfere with efforts to educate citizens to higher levels of scientific literacy. Sagan notes that *A Candle in the Dark* is the title of a book by Thomas Ady, published in London in 1656, attacking the witch-hunts then in progress. Light from the flame of science has illuminated the folly of many superstitions that rely on ignorance of nature's laws and the habits of mind dictated by powerful religious institutions. Sagan identifies self-criticism as one reason for science's success:

> One of the reasons for its success is that science has built-in, error-correcting machinery at its very heart. Some may consider this an overbroad characterization, but to me every time we exercise self-criticism, every time we test our ideas against the outside world, we are doing science. When we are self-indulgent and uncritical, when we confuse hopes and facts, we slide into pseudoscience and superstition. (p. 27).

He goes on to describe the spiritual feeling that can come from understanding nature after considerable effort in thinking about and probing some aspect of our world:

> "Spirit" comes from the Latin word "to breathe." What we breathe is air, which is certainly matter, however thin. Despite usage to the contrary, there is no necessary implication in the word 'spiritual' that we are talking of anything other than matter (including the matter of which the brain is made), or anything outside the realm of science. On occasion, I will feel free to use the word. Science is not only compatible with spirituality; it is a profound source of spirituality. When we recognize our place in an immensity of light-years and in the passages of ages, when we grasp the intricacy, beauty, and subtlety of life, then that soaring feeling, that sense of elation and humility combined, is surely spiritual. (p. 29).

This sounds similar to Einstein's comments about the *cosmic religious feeling* that he associated with a deep appreciation of nature's wonderfully crafted laws. The feeling of spirituality or being intimately connected with nature, loved ones, great humanitarian causes, and so on can be experienced in many ways outside organized, personal-God concepts of religion.

In God We Trust and *God Bless America* are powerful slogans because they play on the strong feelings of spirituality that many associate with the word *God*. The many different God concepts that are represented in the minds of believers all lead to similar feelings of spirituality that can also be experienced by nonbelievers in a variety of other contexts, as Sagan and Einstein suggest. What many think of as cold, heartless science can be as moving and spiritual to the scientist as religious rituals can be to believers in the supernatural or as fly-fishing in a beautiful mountain stream can be to one who loves nature. Many things can trigger the feelings associated with spirituality, just as many things can trigger the feelings associated with fear, anger, guilt, happiness, sexual desire, and other basic human emotions.

Why religious belief is so strong in the United States compared, for example, to many European countries is not easy to determine, but it does help to explain the continued resistance to certain science knowledge, like evolutionary biology. Most European countries have relatively little resistance to evolution education in public schools. Where fundamentalist religious belief is low, tolerance of evolution education, and other science concepts that many in the United States find objectionable, is high.

If one thinks that scientific and religious habits of mind are basically compatible, one's preferred course of action regarding curriculum and instruction in our public schools will be different than it would be if one thinks they are basically incompatible. Because the focus of this book is on science and science education, we will concentrate on these two areas, but the implications for education go well beyond them.

DARWIN'S DANGEROUS IDEA

Ernst Mayr, one of the longest living and many think the most important evolutionary biologist of the twentieth century, considered Charles Darwin the most important scientist who ever lived, and he explained his choice in terms of the depth and breadth of the implications of evolutionary theory for both science and society. Of course it is understandable that an evolutionary biologist would select an evolutionary biologist as the best scientist. However, many others agree that Darwinian evolutionary theory has had immense influence well beyond the borders of science. In *Darwin's Dangerous Idea* (1995), philosopher Daniel Dennett shows how what he calls the *universal acid* of Darwin's great idea penetrates all aspects of our notion of human nature, including ethics and morality. Once considered the private domain of religion, ethics and morality have been invaded by Darwin's dangerous idea. Taking Darwin's idea seriously, not only can it be used to explain the physical and behavioral traits of humans and other animals, it can be extended into the domain of ethics to better understand *why* we behave as we do.

Another important evolutionary biologist who many credit with much of the current interest in sociobiology and evolutionary ethics is Edward Wilson, who, like Mayr, did most of his work while a professor at Harvard University. In an equally important and well-written book, *Consilience: The Unity of Knowledge* (1998), Wilson looks at the implications of taking science, and Darwin's ideas in particular, very seriously. He argues convincingly that those who work in the social sciences and even the humanities can learn something by applying the natural sciences, evolutionary biology in particular, to their studies. In a chapter on ethics and religion Wilson confronts the so-called *naturalistic fallacy* that philosopher G. E. Moore discussed in *Principia Ethica* in 1903. Like philosophers and religionists before and after him, Moore argued that to pass from the factual *is* to the normative *ought* is to commit an error of logic that he called the naturalistic fallacy. Moore placed ethics and moral reasoning in a special category apart from empirical inquiry.

Wilson, Dennett, and others who take natural selection and scientific inquiry seriously follow Darwin's suggestion from *Descent of Man*:

> Thus at last man comes to feel, through acquired and perhaps inherited habit, that it is best for him to obey his more persistent impulses. The imperious word ought seems merely to imply the consciousness of the existence of a rule of conduct, however it may have originated. (p. 486).

Moral reasoning does not have to be placed in a special category. Wilson calls the naturalistic fallacy itself a fallacy and claims that the scientific/empiricist view of ethical reasoning turns traditional, transcendental thinking on its head:

> The empiricist view in contrast, searching for an origin of ethical reasoning that can be objectively studied, reverses the chain of causation. The individual is seen as predisposed biologically to make certain choices. By cultural evolution some of the choices are hardened into precepts, then laws, and if the predisposition or coercion is strong enough, a belief in the command of God or the natural order of the universe... . The empiricist view concedes that moral codes are devised to conform to some drives of human nature and to suppress others. "Ought" is not the translation of human nature but of the public will, which can be made increasingly wise and stable through the understanding of the needs and pitfalls of human nature. (1998, p. 251).

Darwin's dangerous idea, the title of Dennett's 1995 book that describes the far-reaching implications of a serious application of Darwinian evolutionary theory, was used as an introduction by the producers of the outstanding September 2001 PBS/WGBH Boston television series *Evolution: A Journey into Where We're From and Where We're Going*. The eight-hour series concluded with the question *What about God?* For the first time, a major U.S. television production on evolution included the *elephant in the room* issue of religious belief and how it can become an obstacle to an effective science education. Although most of the series was about the history and nature of Darwin's dangerous idea, a significant amount of the production involved the question of God beliefs and how they can interfere with meaningful learning of scientific evolutionary theory. Using Darwin's own struggle with his creationist beliefs during the first two-hour televised program and concluding with accounts of similar struggles by high school and college students in a one-hour finale, *Evolution* explored many of the issues that confront classroom teachers as they try to teach the unifying explanatory theory of biology. It is clear from this that students' early religious training affects their later outlook and habits of mind regarding science concepts that seem to conflict with religious beliefs.

In spite of considerable evidence that creationist religious beliefs are problematic where the learning of Darwinian evolutionary theory is concerned, *Evolution* seemed to favor the more politically correct position that science and religion are compatible. The evidence offered for this position is primarily the fact that some scientists and teachers can do their jobs and still maintain their religious beliefs. The specific religious beliefs of the scientists and teachers in *Evolution* are not explored in any detail, but it seems clear that fundamentalist, creationist beliefs (young Earth, catastrophic flood, and so on) are not part of them. A more figurative, metaphorical interpretation of stories in the Bible or other holy books seems to be the norm for scientists and teachers who suggest that they have no problem being good scientists or teachers while maintaining religious beliefs and practices. However, as the philosophers Martin Mahner and Mario Bunge (1996) point out:

> Liberal religionists tend to see no conflict between science and religion at all. But if their beliefs are supposed to contain at least some true statements about the world, they will finally meet some of the previously listed incompatibilities. After all, the difference between fundamentalist and liberal religion is only a matter of degree, not of kind (p. 110).

Mahner and Bunge go on to explain why they think religious training is an obstacle (although not an insurmountable obstacle) to a science education:

> Yet religious education is an impediment in the sense that it has to be overcome, to be repressed or forgotten, in order to develop a scientific mind. At least the person in question must, at the price of inconsistency, be able to ignore his or her religious metaphysics, value system and attitude in order to do, and while doing, science. All of this is of course possible. But religious education is also an obstacle in the statistical sense that the majority of people are not able to overcome childhood indoctrination. (p. 119)

Mahner and Bunge take a somewhat different approach to the question of compatibility than has been taken throughout this book, but they arrive at the same place, concluding that most religious training is an obstacle to science education. Eliminating religious training is not a viable option in a democracy, where people can choose the nature of their private, religious beliefs and training for their children, so a better science education for students in our public schools seems to be the best option; the remainder of this book focuses on this path. It does seem important, however, to understand the differences and the extent to which there are incompatibilities between scientific and religious habits of mind when considering ways to deal with the elephant in the biology classroom.

DEALING WITH THE ELEPHANT IN THE CLASSROOM

What can be done to help reduce the obstacle of religious belief that many—perhaps most—students in the United States bring to biology classrooms in particular and science classrooms in general? One approach is to ignore the elephant altogether and proceed accordingly. We can call this the *elephant blindness* approach. Another approach is to argue that no real problem exists, so the two are actually compatible. We can call this the *elephant quick-change* approach, giving the elephant any traits we wish. And finally, a third approach is to publicly recognize important differences and incompatibilities between scientific and religious habits of mind and deal with them openly, compassionately, and logically. We can call this the *elephant inquiry* approach.

U.S. court decisions have kept creationism out of most public school biology classrooms, but that does not mean the curriculum is unaffected by the creationist beliefs students and teachers bring to the class. Rather than treat evolution as the central unifying theme that the National Association of Biology Teachers (NABT) says it is, most teachers deal with it as lightly as possible. The PBS production *Evolution* treats evolution as biology's central unifying theme and recognizes that religious beliefs of both students and teachers affect their understanding of and attitude toward this important explanatory theory. In addition to the eight-hour video series, its associated Internet web site (www.pbs.org/evolution) offers a great many resources for teachers and others who want to do a better job helping students understand and more fully appreciate evolution's central role in biology.

In the *elephant inquiry* approach questions arise as to how curriculum and instruction can be changed to facilitate meaningful learning. How should the public school teacher approach this delicate task of confronting certain creationist religious beliefs while staying within the legal guidelines that prohibit the teaching of religion in our public schools?

Taking History Seriously

The Voyage of the Beagle may be the best autobiographical account of the process of conceptual change in biology or even in all of science. It is readable, interesting, and full of many insights into Darwin's gradual process of questioning the traditional creationist viewpoint in response to his experiences in nature and in reading Charles Lyell's *Principles of Geology* during his five-year voyage of discovery. A biology teacher wanting to inject some humanism through history into the biology class could do no better than use material from *The Voyage* with students, preferably early in the school year and in conjunction with the video material on Darwin from the PBS *Evolution* series. The teacher who is knowledgeable about Darwin as a young man before and after the voyage is in a much better position to help students reflect on his long struggle with certain creationist beliefs and how he handled these conflicts. Being conversant with Darwin's *On the Origin of Species* (1859) and *The Descent of Man* (1871), along with *The Voyage*, would place any teacher in a much stronger position to help students better appreciate the nature of evolutionary theory and its implications beyond biology. As the saying goes, we can teach only what we know, and knowledge of the contents of these three books is important for the biology teacher who wants to take history seriously in teaching science. Many secondary sources dealing with Darwin and his ideas and life are available, but one cannot do better than reading the original sources.

Of the many textbooks on evolutionary biology that are available for use by serious high school students and others, one is mentioned here because of its connection to human nature. Other texts are listed in the sources and further reading section that follows. *Biology, Evolution, and Human Nature* (2001), by Timothy Goldsmith (Yale University) and William Zimmerman (Amherst College), is a very well written textbook that lays out the foundations of biology, evolution in action, the biology of behavior, and our place in nature. Using evolution as the central organizing theme, the authors present the key ideas of modern evolutionary science and how they are being used to understand human behavior, and they do it in far fewer pages (about 350) than the typical modern biology textbook.

Confronting the Elephant Directly

Once one recognizes that early religious training can impede the acquisition of a sound science education, where all questions are encouraged and supernatural explanations are excluded, it is important to become aware of frequently asked questions (FAQs) about the relationship between science and religion. Students often ask questions about this relationship, and the effective teacher will be prepared to answer them.

Among these FAQs are:

- Isn't evolution just a theory?
- If we evolved from monkeys, why are there still monkeys around?
- Aren't lots of gaps found in the fossil record?
- How can we be sure evolution occurs if we can't see it happen?
- Is it a sin to believe in evolution?
- Doesn't evolution violate the second law of thermodynamics?
- How can all of the complex forms of life come from chance?
- How can scientists be so confident that Earth is billions of years old?
- Aren't there lots of scientists who reject evolution?
- How can something as complex as DNA originate by chance?
- Aren't scientists just as biased as other people?
- Without God, how can we know what is the right thing to do?

These and other questions about evolution and the nature of science are raised by creationists and many others who simply fail to understand scientific evolutionary theory or how science is conducted. Knowing the facts of scientific evolutionary theory is necessary but not sufficient to ensure adequate answers to these FAQs. Both the facts of evolution theory and the nature of scientific knowledge and how it is produced, checked, revised, and so on are needed in order to adequately understand and explain evolution (or other scientific theories). The U.S. National Academy of Sciences (NAS) recognized the need to help teachers with this and published a guide (*Teaching about Evolution and the Nature of Science*) in 1998. Most of the FAQs in the previous list are dealt with in the guide, as are many other ideas on evolution and the nature of science; the book also includes suggestions for classroom activities for students in biology classes. The NAS recognized the need to integrate ideas on the nature of science into the science of evolution theory rather than handling them separately, as is often done by textbooks. This guide, the PBS video series *Evolution*, and a working knowledge of Darwin's life and work would place the biology teacher in an excellent position to help students develop the kind of science literacy described in *Science for All Americans*.

Confronting the elephant in the room needs to be done with care and compassion, recognizing the importance of students' beliefs and feelings about religion to their overall attitudes regarding evolution and science in general. However, it must be done if we are to be honest and open in our efforts to educate students in the spirit suggested by Dewey in *Democracy and Education*. Religious training at an early age is not done with *Democracy and Education* in mind. Religious training is done to indoctrinate children into accepting certain interpretations of stories in holy books and behaving in ways that leaders in organized religions believe are the preferred ways. To be *Catholic* or *Baptist* or

Methodist or *Hindu* or *Muslim* or *Mormon* or a member of any of the countless variations of these and other organized religious groups means that one must hold certain beliefs, usually about God, and engage in certain rituals that are unique to that particular brand of religion. This *brand loyalty* is taught early and often so that people will continue throughout their lives to support a particular interpretation of God's will. The television preachers try to appeal to all believers, across brands, in an effort to maximize the market share and convince viewers that God needs money from everyone regardless of brand loyalty.

ANSWERING FAQS

For a teacher of science there is no substitute for knowledge of science. We can teach only what we know. Some people suggest that students can construct their own scientific knowledge with little help from a science-knowledgeable teacher. But these people do not understand the unnatural nature of science or how difficult it is to facilitate real conceptual change in students (including adults) when misconceptions are solidly in place. The additional obstacle of religious belief—where evolution is concerned, for example—can make real conceptual change all the more difficult. Simple, one-time, direct answers seldom do much to change the complex cognitive systems that have developed in students' minds over a period of many years. However, understanding science concepts more or less as scientists understand them is required, at a minimum, if teachers hope to be effective in helping students overcome long-held misconceptions about nature's laws.

Whether FAQs are raised by students or teachers, they need to be addressed openly and repeatedly by knowledgeable persons. When a student asks, "Isn't evolution just a theory?" it reflects a misunderstanding of the use of the term *theory* by scientists. Or when a student asks, "Without God, how could we know what is the right thing to do?" it reflects a lack of knowledge of the nature and origin of ethics and morals. Religious belief is not the only place ethics and morals originate, and students should have the opportunity to broaden their perspectives on this and other issues. Great discoveries about nature, like Newton's theory of the universality of gravity, Darwin's theory of natural selection, and Einstein's theory of relativity, unify previously disparate phenomena by showing how they are related. Reducing the origin and diversity of species to the mechanism of natural selection actually unifies many phenomena under one explanatory umbrella, natural selection. Openly considering all questions about nature's laws, including those involving humans and their functioning, is a critical duty of the teacher as professional and a key to an effective education. Dewey's (1916) *Democracy and Education*, Harvey Siegel's (1988) *Educating Reason*, and countless other works on education in a democracy stress the need for openness and critical thought in public education. However, when religious training with young children is effective, it often produces a mind-set that opposes the open, critical environment that characterizes a sound science education.

CONSILIENCE

There are no easy answers to difficult questions in which people are the object of study. The main reason the natural sciences have been so successful in discovering nature's laws is because nature's laws are so stable. Although nature is dynamic (an expanding universe, stars appearing and disappearing, species appearing and disappearing, continents moving and mountains forming, and on and on), its laws and fundamental constituents seem to be remarkably constant. It seems safe to assume, as scientists do, that nature's laws are similar across time and space. The social sciences, on the other hand, are another story. Compared to the natural sciences of biology, chemistry, geology, physics, and so on, the social sciences of anthropology, psychology, sociology, and so on have made little progress. The complexity and variability of humans seems to make the idea of progress in the social sciences a nearly impossible goal. However, during the last few decades that pessimistic outlook has begun to change.

The idea that things in our world, including ourselves, are interconnected is not new, but the idea now can be studied in ways that were not available a few decades ago. One of the remarkable discoveries that makes this possible is the basis of the interconnectedness of all life: the discovery of the similar structure of DNA in all life forms has led scientists, especially evolutionary biologists, and many others to realize that life's fundamental processes are similar. A Howard Hughes Medical Institute report, *The Genes We Share with Yeast, Flies, Worms, and Mice* (2001), provides nontechnical insights into this new conception of the similarities among all living things. This new synthesis is strongly reflected in Wilson's *Consilience*, in which all ways of knowing are seen as interrelated. The Enlightenment period of the seventeenth and eighteenth centuries coincided with the beginnings of modern science (personified by Galileo, Newton, and others), and the ideal of unified knowledge was a common vision among its practitioners. Wilson's *Consilience* tries to regain the ideal of the Enlightenment by placing the natural sciences, biology in particular, at the center of the effort. Evolutionary biology now provides us with the scientific understanding and tools to search for foundations of the social sciences within the natural sciences. Religious belief, for example, can now be examined using scientific methods, and as we saw earlier, neuroscientists are discovering the brain's role in generating the religious mind-set. The electrobiochemical basis of brain function, such as visual perception, psychomotor behavior, and even meditation can be monitored to see where and how the brain functions to form what is called the *mind*.

Opponents of science and of the goals of consilience that Wilson and others support are not restricted to fundamentalists or other defenders of religious tradition. Many academics under the revivalist tent of postmodernism side with fundamentalists and others to oppose the new enlightenment that Wilson describes in *Consilience*. For different reasons, the fundamentalists and many postmodernists oppose the empirical sciences that use scientific methods to make progress in our understanding of the natural world, including humans and their many cultural practices. Just as the creationist movement, since Darwin's *Origin* was published in 1859, has been documented and criticized, so has the postmodern movement during the last few decades—and the criticism from the scientific community has been especially intense.

New disciplines with names like sociobiology, evolutionary psychology, human behavioral genetics, and neuropsychology have appeared at public and private universities and at research institutes, all influenced by advances in science since the 1950s. Wilson summarizes the case for taking science seriously and why pre-scientific minds are so ill-equipped to understand the laws of nature:

> I mean no disrespect when I say that pre-scientific people, regardless of their innate genius, could never guess the nature of physical reality beyond the tiny sphere attainable by unaided common sense. Nothing else ever worked, no exercise from myth, revelation, art, trance, or any other conceivable means; and notwithstanding the emotional satisfaction it gives, mysticism, the strongest pre-scientific probe into the unknown, has yielded zero. No shaman's spell or fast upon a sacred mountain can summon the electromagnetic spectrum. Prophets of the great religions were kept unaware of its existence, not because of a secretive god but because they lacked the hard-won knowledge of physics. (1998, pp. 46–47).

The gradual accumulation of this hard-won knowledge, however tentative at the beginning, adds up to a knowledge base that is solid and without doubt the only reasonably stable explanatory knowledge that humans have been able to achieve. The new disciplines mentioned before suggest that the unification or consilience of the social sciences with the natural sciences has already begun. For example, many psychologists now recognize the importance of taking brain structure and function into account as they try to understand how the mind works, and for leaders in these new fields the distinction between brain and mind has dissolved. The mind is simply the brain in action. Human behavioral geneticists understand the importance of studying other animals as well as humans in the quest to better understand human behavior and related emotions. Very few of these scientists now think that humans are the only animals that experience the range of emotions that, until recently, were reserved for God's special creation. With the important exception of speech, many similarities are noted between two-year-old human infants and their chimpanzee counterparts. And many of Darwin's claims regarding our relationship to gorillas and other apes seem quite unremarkable today, at least to persons informed about primate cognition and behavior.

THE SCIENCE WARS AND POSTMODERNISM

It is not surprising that fundamentalist Christians and other religionists are threatened by evolutionary theory and other scientific theories that are perceived as inconsistent with certain religious beliefs. However, many university academics in the social sciences and humanities have criticized the natural sciences for being too *modern*. Postmodernism developed in the second half of the twentieth century in the United States as a reaction against what some people thought were dominant, repressive ways of thinking and behaving. Science became the target of many postmodernists because science was seen as a dominant, universal form of knowledge that forced other ways of "knowing" into secondary, insignificant roles. Certain feminists criticized science for being too dominated by men, certain racial minorities criticized science for being too dominated by whites, and others criticized science for being the instrument of imperialists. The age of political correctness from the 1960s onward coincided with the academic movement of postmodernism, and during the 1980s and '90s science became the target of those who wanted to show that scientific knowledge was no better than other knowledge. Although few postmodern academics are religious fundamentalists, they find themselves allied with the creationists in criticizing science and scientific knowledge.

Eventually scientists realized that they should respond to some of the criticism coming from the postmodernists; books with titles such as *Higher Superstition: The Academic Left and Its Quarrels with Science*; *A House Built on Sand: Exposing Postmodernist Myths about Science*; and *Fashionable Nonsense: Postmodern Intellectuals' Abuse of Science* began to appear during the mid 1990s. *Fashionable Nonsense* (1998), by physicists Alan Sokol and Jean Bricmont, tells the story of how the so-called Science Wars began and why science's postmodern critics are off the mark when they criticize it for being as subjective and politically sensitive as the social sciences and humanities. A brief look at some of these ideas should help the reader understand why it is not just religionists who criticize science. Habits of mind that are incompatible with scientific habits of mind can and do develop outside religious contexts.

One of the themes that runs through postmodern perspectives is how political and economic power can be used by dominant groups to control others. Exposing the unfair use of power in society is often the focus of postmodern academics, and few argue against this kind of critique unless it requires that evidence and logic be ignored in favor of political correctness.

Religious fundamentalists argue for the inclusion of creationist myths in the public school science curriculum and postmodern academics argue that science should not be seen as superior to other ways of "knowing." All knowledge, argue the postmodern critics of science, is influenced by local cultural factors, including the characteristics of scientists. When taken to the extreme, as many postmodern academics tend to do, this kind of argument leads to what philosophers call *relativism*; since all knowledge is produced by people, all knowledge is equally susceptible to the biases reflected by different religious beliefs, political views, and so on. The slippery slopes of postmodernism/relativism can lead to absurd situations such as viewing astrology and astronomy as equal partners in the search for valid knowledge; in this kind of mental fog religious fundamentalists are quick to claim that their creation myths are as "scientific" as other scientific theories.

It should be noted that not all academics sympathetic to postmodern ideals are anti-science. There are many different views within postmodernism, just as there are many views within the Democratic or Republican parties, or Christianity or Judaism, or Boston Celtics fandom. The characteristic that unites postmodernists is the tendency to reject one idea as being better than another, as this can lead to dominance by one group of people over another. Because science represents the antithesis of this postmodern ideal, many postmodernists oppose the idea that scientific knowledge is better than local or common knowledge about the world. It is this position that I critique here. To expose the vacuous nature of much of the postmodern academics' claims about science, Sokal wrote a paper entitled *Transgressing the Boundaries: Toward a Transformative Hermeneutics of Quantum Gravity* that was full of scientific-sounding nonsense but that used a postmodern vocabulary intended to flatter postmodern academics and convince them that the paper was supportive of their critiques of science. He sent the paper to a top journal in the postmodern academic world and the editors published it, apparently unable to recognize what should have been evident even to college freshmen physics students: that the content itself was intentionally meaningless. The hoax was revealed shortly after the article's publication in mid 1996 and the episode was seen by most observers as further evidence that the postmodern critiques of science, and postmodernism and relativism more generally, should not be taken seriously. Many critiques of postmodernism preceded *Transgressing the Boundaries* and many have appeared since, but none has had the same impact in exposing the postmodern emperor without clothes.

Scientific thought is a threat to authority in society because it requires that independent evidence and reason be supplied to justify the claims of those in power. Cultural tradition is threatened by scientific methods. Scientific ways of thinking can be used to show that claims like *"Women and minorities should be barred from voting and otherwise participating fully in society because they are less capable than men and whites"* are without merit. The Enlightenment's centerpiece was scientific, rational thought, and it continues to be the most valid and reliable way to settle such claims.

In many ways the postmodern relativists are similar to the religious fundamentalists in their criticisms of science. Both groups want to diminish the power of science to judge the truth of claims so that nonscientific claims can enjoy equal (or higher) status. The relativists call their project *multicultural science*, while the fundamentalists call theirs *creation science*. Each group wants to preserve the influence or power enjoyed by traditional parts of the culture. Science, however, is interested in producing ideas that are consistent with nature—and if those ideas are inconsistent with ideas of the past, then so be it.

UNDERSTANDING HUMAN NATURE

The focus throughout this book has been on the nature of scientific and religious habits of mind and how these mind-sets interact to affect human behavior. Religious beliefs and practices vary greatly across different cultures, while science is remarkably similar across those same cultures. Religious belief can be traced back in history for as long as records allow, while science did not really take hold until about four centuries ago. The habits of mind developed through scientific inquiry are very different from the habits of mind developed through most religious training. In comparing these two approaches to understanding our world and ourselves, it is clear that they can differ in nearly every way and, in fact, a strong argument can be made that they are fundamentally incompatible. Yet, it is just as clear that both are powerful influences in our world and will very likely continue to be for some time.

If the last part of the twentieth century is a good indicator, the twenty-first century will change how we think of ourselves—in other words, our human nature—and the science of evolutionary biology that began with Darwin will be at the center of this change. Our public education system will reflect this change by emphasizing more than it ever has the scientific outlook on human nature. As science learns more of the biological basis of human behavior, schools will try to educate our citizens to become more scientifically literate and better prepared to meet life's challenges. A part of this education will include understanding the biological basis of religious beliefs just as we understand the biological basis of other aspects of our human nature. Biology education and social studies education will combine, as Wilson suggests in *Consilience*, to help students better understand all important aspects of human behavior. An indication that this is already beginning to occur is the publication of *Genes, Environment, and Human Behavior* (2000), by Biological Sciences Curriculum Study (BSCS), a longtime leader in developing and publishing cutting-edge curriculum materials for precollege students. This and similar curriculum materials that look at the biological basis of human behavior will encourage teachers of science and social studies to collaborate in efforts to update the school curriculum as science advances our understanding of who we are. In addition, the Human Genome Project and related research will continue to clarify the notion that human nature is inextricably tied up with human evolutionary history.

Interest in evolutionary biology by social scientists continues to increase as more people realize the importance of taking this perspective into account in their respective fields. Even though resistance to evolutionism and sympathy for creationism among many U.S. citizens remains strong, scientists, including many social scientists, understand now more than ever the important role evolutionary perspectives have to play in the social sciences.

Our public education system is of crucial importance in helping children become more open, critical thinkers, and unlike private schools in this country, public schools cannot promote religion and the related habits of mind so carefully cultivated in young children by early religious training. Public schools, unlike private schools and churches, are legally committed to freedom from religious indoctrination, a commitment that follows directly from the First Amendment of the U.S. Constitution. Although private schools and churches may well serve important functions in our society, they are not designed to encourage students to be open, critical thinkers. In fact, many see them as hurdles to these ideals so often associated with democracy. The range of religious beliefs in our society is very large, so it is not accurate to paint all religions or believers with the same brush, but it is fair to say that most religions encourage habits of mind that are inconsistent with scientific habits of mind. This fact has been one of the main points in this book.

Wilson argues in *Consilience* that real progress will be made in the social sciences when the knowledge and methods of the natural sciences are taken seriously. What does it mean to say that progress will be made in religion? Does one kind of religious belief represent progress over another? Most agree now that the religious beliefs associated with witch hunts of the past were bad and we should refrain from such actions in the future. To what extent can we agree on current religious beliefs and practices? Does Wilson's concept of consilience apply to religion as it does to the social and natural sciences? How can the knowledge and methods of the natural sciences be used to make progress in religion? Will a more accurate view of human nature result in "better" religious practice? History suggests that religious beliefs and practices will change as society changes, but the nature of the change, like the evolution of life itself, will be very difficult to predict.

HUMAN NATURE AND EDUCATION

The Mind as Blank Slate

Any theory of education must include ideas about the human mind and how it works. For most of the twentieth century, behaviorist and Freudian ideas dominated American society, including its educational system. Although both of these theories have now been largely discredited by scientists and others who have taken the time to look at them carefully, they continue to exert a strong influence on society. Evolutionary psychologist Steven Pinker uses the metaphor of the blank slate to categorize the various obstacles to a more modern, scientific understanding of human nature and how the mind/brain works. In *The Blank Slate: The Modern Denial of Human Nature* (2002), Pinker argues persuasively for a human nature influenced strongly by evolutionary forces. Until just a few decades ago most people held that it was *nurture* and not *nature* that molded people to become what they are as adults. That is, society or culture was seen as the force that molds children's behavior and directs them into various roles as they become adults.

Many factors, including the eugenics movement during the early 1900s, can be linked to the blank-slate ideology that dominated so much of the social sciences until recently. Trying to control the genes that are passed on to future generations, as was the case in the eugenics movement, poses serious moral questions. From the standpoint of evolutionary biology, our genes are the most important things we possess, and if we are not allowed to pass them to future generations it is the end of the genetic line for us. Other things can be identified as possible causal factors behind blank-slate ideology, including the idea of becoming whatever we want to become in a great democratic country with a tradition of rugged individualism and overcoming all odds to reach our goals. Whatever the causes might have been for the development of the blank-slate ideology, it is clear that nurture rather than nature was dominant in the social sciences for much of the twentieth century. The history of the shift from nurture-only to nature-nurture interaction in human behavior is still being written, but what is of interest here is how current ideas about human nature might influence education in general and science education in particular. If our genes really do influence who we are and what we become, how should this affect our attempts to educate our citizens?

Reconceptualizing education by taking evolutionary science seriously will not be easy. Most people in our society do not understand evolution very well, and many fear it because they have been told by religious leaders that Darwinian evolutionary theory is bad for your spiritual health. Of course spiritual health is defined in many different ways by the various shamans who claim to know what *their* God really wants people to do. Even in religions such as Catholicism, whose pope declared evolution to be factual and consistent with the Catholic faith, many believers still regard evolution with suspicion, and they find it difficult to understand how life's many wonders and great diversity can be explained by natural selection. Evolutionary theory, like so much of science, is often counterintuitive. The various systems that have evolved in the human mind for the purpose of surviving and thriving in our environment are not well adapted to the world of scientific theory. Using this perspective, Pinker observes:

> Education is neither writing on a blank slate nor allowing the child's nobility to come into flower. Rather, education is a technology that tries to make up for what the human mind is innately bad at. Children don't have to go to school to learn to walk, talk, recognize objects, or remember the personalities of their friends, even though these tasks are much harder than reading, adding, or remembering dates in history. They do have to go to school to learn written language, arithmetic, and science, because those bodies of knowledge and skill were invented too recently for any species-wide knack for them to have evolved. (2002, p. 222).

UNLEARNING OLD HABITS OF MIND

One of the most important implications of evolutionary theory for education is that the mind's systems or modules that have evolved over thousands of generations must be understood by educators, and then curriculum and instruction must be adapted for learning things, such as scientific knowledge, that are often inconsistent with how the mind operates. For example, understanding that Earth is about 4,500,000,000 years old and life on Earth is about 3,500,000,000 years old is very difficult to conceptualize, just as the very tiny size and mass of atoms are beyond our concrete grasp. Humans did not need to understand these kinds of things in order to survive, in evolutionary terms, so the mind does not have the kind of machinery needed to easily grasp these concepts. And further complicating the process of understanding many scientific concepts is not just the lack of scientific mind modules, but the interference of real modules the mind does have with the learning of scientific theories. Pinker states the problem nicely:

> Far from being empty receptacles or universal learners, then, children are equipped with a toolbox of implements for reasoning and learning in particular ways, and those implements must be cleverly recruited to master problems for which they were not designed. That requires not just inserting new facts and skills in children's minds but debugging and disabling old ones. Students cannot learn Newtonian physics until they unlearn their intuitive impetus-based physics. They cannot learn modern biology until they unlearn their intuitive biology, which thinks in terms of vital essences. And they cannot learn evolution until they unlearn their intuitive engineering, which attributes design to the intentions of a designer. (2002, p. 223).

It is not simply a matter of piling on more scientific facts about evolution; an effective teacher must recognize that intuitive biology or physics or whatever discipline must be unlearned for meaningful scientific learning to occur. Habits of mind remain after details are a distant memory, and the habit of questioning all hypotheses, regardless of the supposed authority of the hypothesis maker, is at the heart of science.

QUESTIONING COMMON IDEAS ABOUT HUMAN NATURE

1. Sigmund Freud was a great scientist, and his ideas about human nature are important.

Scientists and others who have taken the time to question the validity of Freud's claims about his theory of psychoanalysis have concluded that he was mistaken and very likely knew it from the beginning of his career. Women, in particular, have been harmed by Freudian claims that they are responsible for many of the problematic behaviors of their children. In books with titles like *Seductive Mirage* (Esterson, 1993), *Decline and Fall of the Freudian Empire* (Eysenck, 1986), *The Assault on Truth* (Masson, 1984), *Freudian Fraud* (Torrey, 1992), and *Why Freud Was Wrong* (Webster, 1995), many scholars have shown conclusively that Freud's ideas on human nature were mistaken and that it is likely he knew it early in his career. The vast literature and psychoanalytic practice based on Freud's ideas are, as Webster (1995) says, wrong. The extent to which educational practice, including parenting, is based on Freud's ideas must be recognized and changed to become more consistent with our current best understanding of human nature.

2. Children's basic personality traits are strongly influenced by parental behavior.

In families of more than one child, parents recognize that children develop their own unique personalities even though the home environment is similar for all the children. Studies of identical twins reared apart show that parental influence within a family is relatively small when it comes to basic personality and intelligence traits. In an excellent summary of research on the influence of our genes on basic temperament traits such as shyness, aggressiveness, anger, depression, and addiction, Dean Hamer and Peter Copeland, in *Living with Our Genes* (1998), conclude that our genes are an important part of the causes of these and other aspects of who we are. In other words, our human nature is strongly influenced by our genes, just as other primates' natures are strongly influenced by their genes. It should not be surprising that human behavior, too, like that of our closely related primate cousins, is strongly influenced by genes. Culture, including parental influence, certainly plays an important role in human behavior; however, it is the *interaction* of nature and nurture that produces who we are, not one or the other.

3. Our memories are reliable.

There are many researchers who have shown that human memory can be quite unreliable, but one deserves special mention because of her persistence in the face of great criticism. In *The Myth of Repressed Memory* (Loftus & Ketcham, 1994), Elizabeth Loftus provides compelling evidence that repressed memories of childhood sexual abuse are, for the most part, false. The repressed-memory movement in the United States during the 1980s and early '90s tried to convince the public that many women had repressed memories of sexual abuse during childhood, and for the most part they succeeded. Also, young children in preschools were convinced by therapists (many with Freudian training) that their teachers or caretakers had sexually abused them and that the children had repressed these memories. When provided with the "proper" clues, these children could remember the bad things that these adults had supposedly done to them. What Loftus did as a memory researcher was show that false memories can be implanted in young children by therapists and prosecutors intent on proving that sexual abuse took place when in fact it did not. She also showed that adults are easily convinced that they have repressed memories of events in their childhood that never took place. Even eyewitness accounts of events are often quite unreliable, with different people giving conflicting accounts of the same event. Human memory is remarkable for its ability to store a huge amount of information, but unfortunately it is prone to many kinds of errors.

4. What our senses and mind tell us about the world is true.

The sensory data provided by our eyes, ears, and so on are very important and usually very good at keeping us alive, but they are not good at providing scientific accounts of nature. To a large extent Earth does seem flat, and the sun, planets, and stars do seem to revolve around the Earth, just as our moon does. It is difficult to believe that continents move and mountains are formed by the movement of huge landmasses. It seems incredible that all of the instructions for all life are encoded in microscopic molecules structured as a double helix and that our emotions and feelings can be traced to various chemicals in certain parts of our brains. The history of science is a history of conflict with our commonsense ideas about the world. Our senses and mind simply do not operate in ways that make it easy for us to understand nature's laws. In fact, as British biologist Lewis Wolpert suggests, our commonsense ideas about the world lead us to misunderstandings about nature. We simply cannot trust our senses to reveal the truth about nature's laws. This aspect of our human nature, more than any other, is the most problematic when it comes to helping children and adults understand scientific knowledge. This is where we must concentrate our efforts if we hope to provide our children with a sound science education.

5. At birth the mind is a blank slate.

The blank-slate doctrine assumes that reality can be inserted or written into the mind. Another metaphor that is used for this doctrine is the mind as *empty vessel*. Because a baby seems to have no experience with the world it is assumed by blank-slate enthusiasts that it has no preexisting tendencies or preferences. To properly educate a child one can simply provide direct instruction about the desired knowledge and the mind automatically responds by memorizing the information, without regard to preexisting mind/brain biases. This incorrect assumption about human nature is thoroughly criticized by Pinker in *The Blank Slate* (2002). Although a child's environment can be very influential in shaping who the child becomes, there are many preexisting, genetic tendencies that interact with environmental forces to shape the developing child. It is the job of the parent, teacher, and others who hope to help a child become a thoughtful, critical, educated adult to know about preexisting biases of the mind/brain and then teach accordingly.

6. The mind and the brain are very different things.

The understanding that the mind is simply the brain in action did not come into focus until the second half of the twentieth century, when cognitive science, neuroscience, and evolutionary biology provided strong evidence that body and mind are one and the same thing. The recognition that information processing is what the brain does as the body's senses provide data was a very important advance in understanding the mind/brain. There are universal mental mechanisms or modules that are characteristic of all people regardless of cultural differences. For example, even though each culture may have its own language, there are fundamental similarities in the grammar of all languages. These universal patterns in language usage suggest underlying similarities in the minds/brains of all people; most linguists who take biology seriously agree that a language module has evolved in the same sense that other organs in the human body have evolved. Just as the body is composed of many interacting parts, so the mind/brain is composed of many interacting modules, each with a special job to do, and evolution is the ultimate explanation behind the nature of all the bodies' many interacting parts. Just as the idea of natural selection eliminated the need for a supernatural cause behind the evolution of new species, mind-as-brain-functioning eliminates the need for mysterious causes to explain how we are able to confront and solve life's many problems. Pinker's *How the Mind Works* (1997) provides a nice summary of this new perspective of the mind/brain.

7. We can become whatever we choose to become.

The U.S. Constitution and Bill of Rights and other amendments state that all citizens should have equal opportunities under the various laws of the land. Many people also believe that through free will we can become whatever we choose to become, given enough hard work and a few lucky breaks. The idea that our genetic makeup may play an important part in influencing who or what we become is dismissed by the free-will advocates as sociobiology baloney. If we want to become a heterosexual, right-handed mathematician, all we have to do is commit to it, work hard, and eventually it will happen. Free will can conquer all. The problem is that sexual orientation and handedness are heavily influenced by our genetic makeup and even certain mathematical prowess seems to be influenced considerably by things other than the opportunity to learn mathematics. No matter how hard we try, if we are born with left-handed tendencies we will prefer to use our left hand in most situations and if we are born with homosexual tendencies we will prefer to have sexual encounters with persons of the same sex. It is less clear that mathematical fluency is so strongly influenced by biological factors, but even here it seems clear that mathematical genius is more than simply wanting to be good at mathematics. The will to succeed is itself influenced by our biology, as we see with people who suffer from depression and anxiety disorders. The free-will advocates must ignore a great deal of solid science as well as much anecdotal evidence to maintain that we can become whatever we choose to become.

Human nature must be taken seriously by parents, teachers, and others who are interested in the education of our children and, for that matter, in all education that seeks to help people become independent, critical, rational thinkers who enjoy learning about their world. For parents and teachers to ignore the commonalities of the mind/brain across cultures in the twenty-first century is to act irresponsibly, for it is in these similarities that solid principles of teaching and learning will be grounded.

WHERE DO WE GO NOW?

The spiritual, religious side of people seems to be a part of our human nature, while the rational, scientific side seems alien to most. Belief in the supernatural is so strong for some that they are willing to give up their life for what they believe their God wants them to do. Some scholars think that the source of religious belief in unseen forces is fear. We fear many things in life, including the death of our loved ones and ourselves. Fear is a natural, powerful emotion that helps us stay alive in dangerous times, so inventing unseen forces that can protect us in dangerous, difficult times might have been beneficial to early believers, helping them overcome life's challenges. Whatever the early history of religious belief might be, it is clear that most people in the United States believe in a personal God who watches over them and has the power to punish and reward. And it is just as clear that, unlike science, the existence or nature of God cannot be determined by logic and analysis of evidence. One must simply believe in these unseen forces and the stories found in holy books and told in places of worship, usually starting as a young child and continuing into adulthood.

The kind of spirituality that Einstein said was behind his search for nature's laws is uncommon among believers. His rejection of a personal God, who rewards and punishes and suspends nature's laws through miracles, in favor of an impersonal Supreme Being behind nature's laws does not seem to appeal to many believers. Most people seem to want a God who is more personal and in charge of things, a bit like a powerful king who sits in a castle on a hilltop overlooking his kingdom and communicates with certain people in the kingdom, who then share these messages with the masses. Because the hilltop is surrounded by clouds, most people never actually see the king and his castle but they are taught at an early age to believe what the messengers tell them about the king's wishes and commands. The mere act of disbelief or doubt can be enough to anger the king and his messengers and cause bad things to happen to the doubter or his family and friends.

The abstract, often counterintuitive world of science, with its mathematics and experiments and measurements and logic, is not seen as very user-friendly by most people, so they try to avoid it. Developing the habits of mind associated with scientific thought, like questioning authority, being skeptical of claims for which the evidence is thin or nonexistent, placing high value on the results of careful investigation, and rejecting supernatural causes of natural phenomena, is a real challenge. The twentieth century can easily be called the century of science, but few people in the twenty-first century are considered scientifically literate by educational commissions charged with keeping track of such things. The habits of mind associated with pseudoscience and superstition are still widespread: people today seem to be just as easily convinced of things for which there is no credible evidence, such as astrology and numerology and aliens in flying saucers and angels and devils and psychics who communicate with the dead, and on and on, as were those of much earlier generations. The lack of skepticism toward such phenomena might not be surprising in pre-scientific cultures prior to the twentieth century, but after more than a century of modern science it is surprising to many scholars that pre-scientific thought is still so widespread among people in countries like the United States. It is not unreasonable to conclude that humans have evolved in such a way that makes pseudoscience and magic and mysticism seem like natural, acceptable ways of believing and behaving. Believing or wanting to believe in things for which there is no credible evidence seems to be a universal trait among all cultures, just as there are many other human universals such as baby talk, childhood fears, facial expressions, marriage, music, myths, fear of death, and so on. What Carl Sagan called the demon-haunted world (1996) seems to be nearly as characteristic of the twenty-first-century world as of that of prior centuries. The sixty-four-dollar question is *How can we get people to become more scientific in their approach to life's important problems and challenges?* Or, in other words, *How can we reduce people's tendency to blindly accept authority and claims for which there is little or no credible evidence?*

There are no easy answers to these questions, but an effective science education, including solid research related to teaching and learning, is the choice of many scholars who have carefully studied the problem. In keeping with the focus of this book, the nature and interaction of scientific and religious habits of mind, I offer some closing thoughts and ideas for action.

1. Understand that certain habits of mind that are characteristic of certain kinds of religious training can hamper efforts toward a sound science education.

Although more research is needed to explore the nature of the interaction of scientific and religious habits of mind, it is clear that open inquiry is hampered when certain assumptions about our world are made. Nature's laws are discovered through open inquiry, not through uncritical acceptance of religious, political, or other dogma. The history of science is a history of resisting ideas because they seem to conflict with long-held assumptions, often religious in origin, about the nature of our world. Once it is recognized that a sound science education can be hampered by habits of mind such as acceptance of claims without solid evidence, those who value scientific literacy will be in a better position to take effective action.

2. Understand that organized religion and belief in a personal God are deeply embedded in U.S. society and should not be treated lightly.

It is just as clear that most U.S. citizens have strong beliefs in a personal God, and these beliefs must be allowed as long as they do not interfere with the rights of others, as protected by the Constitution and its various amendments. Freedom of religion and freedom from religion are equally protected in our democracy. The separation of church and state is threatened from time to time by politicians, religious leaders, and others who misunderstand the meaning of the separation clause in the first amendment—"Congress shall make no law respecting an establishment of religion." Regardless of what many politicians and others suggest, these ten words mean what they say: keep religion out of the laws of the land. Personal religious beliefs must not be allowed to govern what others want to believe about the supernatural. Public schools must refrain from promoting any particular religious beliefs. The thousands of different God beliefs held by the many different religious sects in the United States are not to be promoted or inhibited by the public schools or other public institutions, regardless of whether politicians and other public officials use the phrase *God Bless America* when they think it will benefit them in some way.

3. Understand that science, unlike religion, is not a natural way of thinking or behaving and when the two seem to conflict, as with creationism and evolution, science often loses in the mind of the believer.

Because so much of science is counterintuitive, it will not happen easily or naturally that most people will become scientifically literate. Many prescientific conceptions about our world must be overcome before scientific thought becomes natural; the history of science provides many clues regarding these conceptions. The science teacher who understands the history and nature of her particular field of science will be in a better position to help students overcome misconceptions that, at one time in history, all persons had to overcome. When scientific knowledge seems to conflict with religious belief, history shows that religious leaders eventually must change their claims about reality. However, this often takes a long time, and children in schools are caught in the middle of conflicting truth claims and will often follow the lead of parents and pastors and peers rather than science textbook authors and teachers. The young-Earth creationist is seldom moved by evidence derived from radiometric dating techniques that Earth is over 4.5 billion years old and life on Earth is over 3.5 billion years old. Understanding something of the current scientific ideas about how the mind works can be helpful to the teacher who wants to help students overcome pre-scientific conceptions of the world.

4. Understand that good science teaching must include an emphasis on the nature of scientific inquiry.

Students must understand how scientific knowledge is generated if they are to be considered scientifically literate. It is not enough to know the facts and theories of science, although this knowledge is a necessary part of scientific literacy. How students learn the concepts and theories of science will determine, to a large extent, what they understand of the nature of scientific inquiry. Investigations that test hypotheses, excluding some and allowing others to stand for at least a while, can show the importance of hands-on inquiry and help students see the difference between scientific and nonscientific approaches to testing different claims. Collecting and analyzing real data in an effort to make decisions about competing hypotheses or simply to learn more about some aspect of nature before hypotheses are well formed, should be a central part of any science education program.

5. Understand that open inquiry is a central tenet of both democracy and science.

Openness is crucial to democratic forms of government and to progress in science. Citizens must be free to inquire into the workings of government and to report and criticize what they find just as scientists must be free to question and criticize the work of their colleagues. Knowledge is preferred over ignorance, even though knowledge can sometimes be disconcerting when it conflicts with preexisting beliefs. Questioning authority is a necessary part of both a healthy democracy and a science that hopes to make progress beyond the current state of knowledge. Debate and argumentation, using evidence and logic, are important in both science and in democratic forms of government, so disagreements should not be seen as undesirable but rather as a normal part of the process of maintaining a healthy democracy or a progressive science.

6. Understand that religious beliefs vary widely even among persons of the same church, temple, mosque, or other places of worship.

The history of religion is a history of widely varying belief systems. From multiple gods and goddesses to single gods, people have maintained that their particular unseen forces are superior to their neighbor's and they are often willing to kill and wage war to prove it. Each god or goddess is crafted according to the wishes of those in a culture who claim to have special insight into the nature and desires of these supreme beings, and certain rituals are followed in order to please the gods. If students could learn in history of religion courses or elsewhere that human history is marked by widely varying religious beliefs, continuing to the present day, and that their particular religious beliefs are a result of their particular life experiences, especially their parents' religious preferences, greater religious tolerance might settle into society. Of course it probably depends on how tolerant students are taught to be during their early years. Tolerance to differences in religious beliefs seems to be a difficult thing to teach, just as scientific literacy seems to be a difficult thing to achieve. Perhaps there is a connection here that needs further investigation.

7. Understand the nature and history of scientific inquiry and how it varies across the natural sciences.

Scientific inquiry varies across the natural sciences. The inquiry methods used by evolutionary biologists differ from the methods used by nuclear physicists, for example, even though both must build their theories on the reality of nature. The inquiry methods of Isaac Newton, including the mathematics available over three centuries ago, were different from the methods used by Albert Einstein during the early twentieth century, and both are considered great physicists. Scientists use whatever methods are available at a given time to learn more of nature's laws, and when new methods become available new discoveries often are made. The common goal of better understanding the laws of nature is accomplished using whatever methods seem best suited for the job at the time, and when the results of an investigation can be replicated by many scientists the scientific community usually accepts the validity of the results. Unlike religious dogma, however, all scientific knowledge must remain open to challenge by future researchers.

8. Understand that the nature of the interaction of religious training and a scientific education can and should be studied using scientific methods.

Very little research has been done to explore the relationship between early religious training and a later science education. Do certain religious beliefs interfere with one's ability to acquire scientific habits of mind? Can a person believe that Earth is only a few thousand years old and learn in a meaningful way that Earth is a few billion years old? Can a person believe that God created everything in a few days and learn in a meaningful way that life on Earth evolved over billions of years? How are conflicts such as these resolved and what is their impact on the development of scientific habits of mind? When teachers have conflicts like this in their own minds how is their teaching affected?

From the large body of research on misconceptions we know that people have difficulty acquiring scientific conceptions of nature. Conceptual change in the face of tightly held misconceptions is often a long and difficult process, whether or not the source of the misconception is religious belief. It is difficult enough to try to understand the scientific theory of evolution, for example, without the additional problem of believing it is wrong to do so.

A little over a decade ago a conference on evolution education research (Good et al., 1993), funded by the National Science Foundation and held at Louisiana State University, identified research needed in evolution education. Scientists, science teachers, and science teacher educators worked to develop a research agenda to explore the difficulties students have in learning scientific evolutionary theory, but the issue of religious beliefs was avoided. It is now clear to this author that religious beliefs/habits of mind are often related to pre-scientific ideas about nature's laws and that more research in this area is needed. If we hope to better understand how to increase the scientific literacy and related habits of mind of citizens we must confront the reality that religious beliefs often act as obstacles. Just as that conference at LSU helped to direct attention toward the need for more and better research on evolution education, I hope this book helps to increase awareness of the need for research on the connection between scientific and religious habits of mind.

SOURCES AND FURTHER READING

Aczel, A. 1999. *God's equation: Einstein, relativity, and the expanding universe.* New York: Delta.

Alcock, J. 2001. *The triumph of sociobiology.* New York: Oxford University Press.

Aleixandre, M. 1994. Teaching evolution and natural selection: A look at textbooks and teachers. *Journal of Research in Science Teaching* 31: 519–35.

Alexander, R. 1987. *The biology of moral systems.* Hawthorne, NY: Aldine.

Allman, J. 1999. *Evolving brains.* New York: Scientific American.

Alper, M. 2000. *The God part of the brain.* New York: Rogue Press.

Alters, B., & S. Alters. 2001. *Defending evolution: A guide to the creation/evolution controversy.* Sudbury, MA: Jones and Bartlett.

American Association for the Advancement of Science. 1989. *Science for all Americans.* New York: Author.

———. 1993. *Benchmarks for science literacy.* New York: Author.

Armstrong, K. 1993. *A history of God: The 4,000-year quest of Judaism, Christianity, and Islam.* New York: Basic Books.

Arnhart, L. 1998. *Darwinian natural right: The biological ethics of human nature.* Albany, NY: SUNY Press.

Austin, J. 1998. *Zen and the brain: Toward an understanding of meditation and consciousness.* Cambridge, MA: MIT Press.

Bachelard, Gaston. 1984. *The new scientific spirit.* Boston: Beacon. (Originally published 1934.)

Bagemihl, B. 1999. *Biological exuberance: Animal homosexuality and natural diversity.* New York: St. Martin's.

Barkow, J., L. Cosmides, & J. Tooby. 1992. *The adapted mind: Evolutionary psychology and the generation of culture.* New York: Oxford University Press.

Barlow, N., ed. 1969. *The autobiography of Charles Darwin, 1809–1882.* New York: Norton. (Original work published 1888.)

Barondes, S. 1999. *Molecules and mental illness.* New York: Scientific American Library.

Barton, R. 1997. *Visual specialization, brain evolution, and behavioural ecology in primates.* Berlin: Blackwell.

Betzik, L., ed. 1997. *Human nature: A critical reader.* New York: Oxford University Press.

Bevilacqua, F., E. Gianetto, & M. Matthews, eds. 2001. *Science education and culture: The contribution of history and philosophy of science.* Boston: Kluwer.

Biological Sciences Curriculum Study. 2000. *Genes, environment, and human behavior.* Colorado Springs, CO: Author.

Bishop, B., & C. Anderson. 1990. Student conceptions of natural selection and its role in evolution. *Journal of Research in Science Teaching* 27: 415–27.

Bishop, G. 1998. What Americans really believe. *Free Inquiry* 19 (summer): 38–41.

Blum, D. 1998. *Sex on the brain: The biological differences between men and women.* New York: Penguin.

Boyd, R., & P. Richerson. 1985. *Culture and the evolutionary process.* Chicago: University of Chicago Press.

Boyd, R., & L. Silk. 1997. *How humans evolved.* New York: Norton.

Boyer, P. 1994. *The naturalness of religious ideas: A cognitive theory of religion.* Berkeley: University of California.

———. 2001. *Religion explained: The evolutionary origins of religious thought.* New York: Basic Books.

Brooke, J. 1991. *Science and religion: Some historical perspectives.* New York: Cambridge University.

Brown, D. 1991. *Human universals.* New York: McGraw-Hill.

Bruer, J. 1999. *The myth of the first three years: A new understanding of early brain development and life-long learning.* New York: Free Press.

Bryant, P. 1992. Arithmetic in the cradle. *Nature* 358: 712–13.

Burkert, W. 1996. *Creation of the sacred: Tracks of biology in early religion.* Cambridge, MA: Harvard University Press.

Burr, C. 1996. *A separate creation: The search for the biological origins of sexual orientation.* New York: Hyperion.

Buss, D. 1994. *The evolution of desire.* New York: Basic Books.

———. 1999. *Evolutionary psychology: The new science of the mind.* Needham Heights, MA: Allyn & Bacon.

Byrne, R. 1995. *The thinking ape: Evolutionary origins of intelligence.* New York: Oxford University Press.

Carey, S. 1998. Knowledge of number: Its evolution and ontogeny. *Science* 282: 641–42.

Chomsky, N. 1998. *Language and the problems of knowledge.* Cambridge, MA: MIT Press.

———. 2000. *New horizons in the study of language and mind.* New York: Cambridge University Press.

Clark, R. 1971. *Einstein: The life and times.* New York: Avon.

———. 1976. *The life of Bertrand Russell.* New York: Knopf.

Colapinto, J. 2000. *As nature made him: The boy who was raised as a girl.* New York: HarperCollins.

Collins, W., E. Maccoby, L. Steinberg, E. Hetherington, & M. Bornstein. 2000. Contemporary research on parenting: The case for nature and nurture. *American Psychologist* 55: 218–32.

Crick, F. 1994. *The astonishing hypothesis: The scientific search for the soul.* New York: Simon & Schuster.

Cromer, A. 1993. *Uncommon sense: The heretical nature of science.* New York: Oxford University Press.

Cushing, J. 1995. *Quantum mechanics: Historical contigency and the Copenhagen Hegemony.* Chicago: University of Chicago Press.

Daly, M., & M. Wilson. 1999. *The truth about Cinderella: A Darwinian view of parental love.* New Haven, CT: Yale University Press.

Damasio, A. 1999. *The feeling of what happens: Body and emotion in the making of consciousness.* New York: Harcourt Brace.

D'Aquili, E. 1999. *The mystical mind: Probing the biology of religious experience.* Minneapolis, MN: Fortress.

Darwin, C. 1962. *The voyage of the Beagle,* ed. Leonard Engel. New York: Doubleday. (Original 1839 work published 1860.)

———. 1859. *On the origin of species.* London: Murray.

———. 1871. *The descent of man, and selection in relation to sex.* London: Murray.

———. 1872. *The expression of emotions in man and animals.* Oxford: Oxford University Press.

Dawkins, R. 1976. *The selfish gene.* Oxford: Oxford University Press.

———. 1987. *The blind watchmaker: Why the evidence of evolution reveals a universe without design.* New York: Norton.

———. 1998. *Unweaving the rainbow: Science, delusion, and the appetite for wonder.* Boston: Houghton Mifflin.

Deacon, T. 1997. *The symbolic species: The coevolution of language and the brain.* New York: Norton.

Dehaene, S. 1997. *The number sense: How the mind creates mathematics.* New York: Oxford University Press.

Demastes/Southerland, S., R. Good, & P. Peebles. 1996. Patterns of conceptual change in evolution. *Journal of Research in Science Teaching* 33: 407–31.

Dennett, D. 1995. *Darwin's dangerous idea: Evolution and the meanings of life.* New York: Simon & Schuster.

Desmond, A., & J. Moore. 1992. *Darwin: Life of a tormented evolutionist.* New York: Touchstone.

de Waal, F. 1996. *Good natured: The origins of right and wrong in other animals.* Cambridge, MA: Harvard University Press.

Dewey, J. 1916. *Democracy and education.* New York: Macmillan.

Diamond, J. 1997. *Why is sex fun? The evolution of human sexuality.* New York: Basic Books.

Dobzhansky, T. 1973. Nothing in biology makes sense except in the light of evolution. *American Biology Teacher* 35: 125–29.

Donald, M. 1991. *Origins of the modern mind: Three stages in the evolution of culture and cognition.* Cambridge, MA: Harvard University Press.

Dorit, R. 1997. Review of Michael Behe's "Darwin's black box." *American Scientist* 85: 474–75.

Dugatkin, L. 1997. The evolution of cooperation. *Bioscience* 47: 355–62.

Durham, W. 1991. *Coevolution: Genes, culture, and human diversity.* Stanford, CA: Stanford University Press.

Duveen, J., & J. Solomon. 1994. The great evolution trial: Use of role-play in the classroom. *Journal of Research in Science Teaching* 31: 575–82.

Ehrlich, P. 2000. *Human natures: Genes, cultures, and the human prospect.* Washington, DC: Island.

Einstein, A. 1924. *The theory of relativity.* London: Constable.

———. 1950. *Out of my later years.* London: Constable.

———. 1982. *Ideas and opinions.* New York: Crown. (Originally published 1954.)

Ekman, P., & R. Davidson, eds. 1994. *The nature of emotion: Fundamental questions.* New York: Oxford University Press.

Eldridge, N. 2000. *The triumph of evolution and the failure of creationism.* New York: Freeman.

Eysenck, H. 1986. *Decline and fall of the Freudian empire.* New York: Viking.

Esterson, A. 1993. *Seductive mirage: An exploration of the work of Sigmund Freud.* Chicago: Open Court.

Feldman, M. 1997. Twin studies, heritability, and intelligence. *Science* 284: 278.

Filkin, D. 1997. *Stephen Hawking's universe: The cosmos explained.* New York: Basic Books.

Freeman, D. 1999. *The fateful hoaxing of Margaret Mead: A historical analysis of her Samoan research.* Boulder, CO: Westview.

Frith, C., & U. Frith. 1999. Interacting minds: A biological basis. *Science* 286: 1692–95.

Foley, R. 1995. *Humans before humanity.* Oxford: Blackwell.

Fortey, R. 1997. *Life: A natural history of the first four billion years of life on Earth.* New York: Knopf.

Futuyma, D. 1995. *Science on trial: The case for evolution.* Sunderland, MA: Sinauer.

———.1998. *Evolutionary biology.* Sunderland, MA: Sinauer.

Gallup, G., & F. Newport. 1991. Belief in paranormal phenomena among adult Americans. *Skeptical Inquirer* 15: 137–46.

Gardner, H. 1987. *The mind's new science: A history of the cognitive revolution.* New York: Basic Books.

Gardner, M. 1991. *The new age: Notes of a fringe watcher.* New York: Prometheus.

Gazzaniga, M., ed. 2000. *The new cognitive neurosciences.* Cambridge, MA: MIT Press.

Geary, D. 1995. Reflections on evolution and culture in children's cognition. *American Psychologist* 50: 24–37.

Gellner, E. 1992. *Postmodernism, reason, and religion.* New York: Routledge.

Gibbons, A. 1998. Which of our genes make us human? *Science* 281: 1432–34.

Goldsmith, T. 1990. Optimization, constraint, and history in the evolution of eyes. *Quarterly Review of Biology* 65: 281–322.

Goldsmith, T., & W. Zimmerman. 2001. *Biology, evolution, and human nature.* New York: Wiley.

Good, R., M. Hafner, & P. Peebles. 2000. Scientific understanding of sexual orientation: Implications for education. *American Biology Teacher* 62: 326–30.

Good, R., & J. Shymansky. 2001. Nature of science literacy in *Benchmarks* and *Standards*: Post-modern/relativist or modern/realist? In *Science education and culture*, ed. F. Bevilacqua, E. Giannetto, & M. Matthews. New York: Kluwer.

Good, R., J. Shymansky, & L. Yore. 1999. Censorship in science and science education. In *Caught off guard: Teachers rethinking censorship and controversy*, ed. E. Brinkley. Boston: Allyn and Bacon.

Good, R., J. Trowbridge, S. Demastes/Southerland, J. Wandersee, M. Hafner, & C. Cummins. 1993. *Proceedings of the 1992 evolution education research conference.* Baton Rouge, LA: Authors.

Good, R., & J. Wandersee. 1997. Cognitive travels via the H.M.S. *Beagle*: A study of Darwin's voyage of conceptual change and its implications for teaching about evolution. In *Proceedings of the Fourth International Conference on History & Philosophy of Science in Science Teaching*, ed. L. Lentz & I. Winchester. Calgary, June 21–25.

Gould, S. 1977. *Ever since Darwin.* New York: Norton.

Griffin, D. 1992. *Animal minds.* Chicago: University of Chicago Press.

Gross, P., N. Levitt, & M. Lewis, eds. 1996. *The flight from science and reason.* Baltimore, MD: Johns Hopkins University Press.

Hamer, D., & P. Copeland. 1998. *Living with our genes.* New York: Anchor.

Hamilton, W. 1964. The genetic evolution of social behavior. *Journal of Theoretical Biology* 7: 1–52.

Harris, J. 1998. *The nurture assumption: Why children turn out the way they do.* New York: Free Press.

Haught, J. 2000. *God after Darwin: A Theology of Evolution.* Boulder, CO: Westview.

Hirschfeld, L., & S. Gelman. 1994. *Mapping the mind: Domain specificity in cognition and culture.* New York: Cambridge University Press.

Holton, G. 1988. *Thematic origins of scientific thought: Kepler to Einstein.* Cambridge, MA: Harvard University Press.

———. 1993. *Science and anti-science.* Cambridge, MA: Harvard University Press.

———. 1996. *Einstein, history, and other passions.* New York: Addison-Wesley.

Howard Hughes Medical Institute. 2001. *The genes we share with yeast, flies, worms, and mice.* Chevy Chase, MD: author

Hull, D. 1988. *Science as a process: An evolutionary account of the social and conceptual development of science.* Chicago: Chicago University Press.

Jammer, Max. 1999. *Einstein and religion.* Princeton, NJ: Princeton University Press.

Joseph, R. 2000. *The transmitter to God: The limbic system, the soul, and spirituality.* San Jose, CA: University of California Press.

Kagan, J. 1998. *Three seductive ideas.* Cambridge, MA: Harvard University Press.

Kalat, J. 1998. *Biological Psychology.* Pacific Grove, CA: Brooks/Cole.

Keynes, R. 2001. *Darwin, his daughter, and human evolution.* New York: Riverhead.

Kitcher, P. 1996. *The lives to come: The genetic revolution and human possibilities.* London: Penguin.

Klein, R. 1999. *The human career: Human biological and cultural origins.* Chicago: University of Chicago Press.

Koertge, N., ed. 1998. *A house built on sand: Exposing postmodernist myths about science.* New York: Oxford University Press.

Kohn, D. 1989. Darwin's ambiguity: The secularization of biological meaning. *British Journal for the History of Science,* 22, 215–39

Kragh, H. 1999. *Quantum generations: A history of physics in the twentieth century.* Princeton, NJ: Princeton University Press.

Kurtz, P. 1992. *The new skepticism: Inquiry and reliable knowledge.* Buffalo, NY: Prometheus Books.

Lakoff, G., & R. Nunez. 2000. *Where mathematics comes from: How the embodied mind brings mathematics into being.* New York: Basic Books.

Larson, E. 1997. *Summer for the Gods: The Scopes trial and America's continuing debate over science and religion.* Cambridge, MA: MIT Press.

Lawson, A. 1999. A scientific approach to teaching about evolution and special creation. *The American Biology Teacher* 61: 266–74.

Lawson, A., & W. Worsnop. 1992. Learning about evolution and rejecting a belief in special creation: Effects of reflective reasoning skill, prior knowledge, prior belief, and religious commitment. *Journal of Research in Science Teaching* 29: 143–66.

LeVay, S. 1996. *Queer science.* Cambridge, MA: MIT Press.

LeVay, S., & D. Hamer. 1994. Evidence for a biological influence in male homosexuality. *Scientific American* 270: 44–49.

Li, W. 1997. *Molecular evolution.* Sunderland, MA: Sinauer.

Lieberman, P. 1998. *Eve spoke: Human language and human evolution.* New York: Norton.

Loftus, E., & K. Ketcham. 1994. *The myth of repressed memory: False memories and allegations of sexual abuse.* New York: St. Martin's Griffin.

Lumsden, C., & E. Wilson. 1981. *Genes, mind, and culture.* Cambridge, MA: Harvard University Press.

Lyell, C. 1830–33. *Principles of geology.* 3 volumes. London: Murray.

Mahner, M., & M. Bunge. 1996. Is religious education compatible with science education? *Science & Education* 5: 101–23.

Malthus, T. 1826. *An essay on the principle of population.* 6th ed. London: Murray.

Martin, M. 1997. Is Christian education compatible with science education? *Science & Education* 6: 239–49.

Masson, J. 1984. *The assault on truth: Freud's Suppression of the seduction theory.* New York: Farrar, Straus and Giroux.

Matthews, M. 1994. *Science teaching: The role of history and philosophy of science.* New York: Routledge.

Mayr, E. 1997. *This is biology: The science of the living world.* Cambridge, MA: Harvard University Press.

———. 1991. *One long argument: Charles Darwin and the genesis of modern evolutionary thought.* Cambridge, MA: Harvard University Press.

McGuire, M., & A. Troisi. 1998. *Darwinian psychiatry.* New York: Oxford University Press.

Miller, K. 1999. *Finding Darwin's God: A scientist's search for common ground between God and evolution.* New York: Perennial.

Mintzes, J., J. Wandersee, & J. Novak, eds. 1997. *Teaching science for understanding: A human constructivist view.* New York: Academic.

Mivart, G. 1859. *On the genesis of species.* London: Macmillan.

Nanda, M. 1997. The science wars in India. *Dissent* (winter): 78–83.

National Academy of Sciences. 1998. *Teaching about evolution and the nature of science.* Washington, DC: National Academy Press.

National Research Council. 1996. *National science education standards.* Washington, DC: National Academy Press.

Newberg, A., E. D'Aquili, & V. Rause. 2001. *Why God won't go away: Brain science and the biology of belief.* New York: Ballantine.

Newton, R. 1997. *The truth of science: Physical theories and reality.* Cambridge, MA: Harvard University Press.

Newton-Smith, W. 1981. *The rationality of science.* New York: Routledge.

Nitecki, M., & D. Nitecki, eds. 1993. *Evolutionary ethics.* Albany NY: SUNY Press.

Norris, C. 1997. *Against relativism: Philosophy of science, deconstruction, and critical theory.* Malden, MA: Blackwell.

Numbers, R. 1992. *The creationists: The evolution of scientific creationism.* Berkeley: University of California Press.

Paradis, J., & G. Williams. 1989. *Evolution and ethics.* Princeton, NJ: Princeton University Press.

Parker, S., & M. McKinney. 1999. *Origins of intelligence: The evolution of cognitive development in monkeys, apes, and humans.* Baltimore, MD: Johns Hopkins University Press.

Pennisi, E. 1999. Are our primate cousins conscious? *Science* 284: 1984–85.

Pennock, E. 1999. *Tower of Babel: The evidence against the new creationism.* Cambridge, MA: MIT Press.

Persinger, M. 1987. *Neuropsychological bases of God beliefs.* New York: Praeger

Persinger, M. 1993. Vectorial cerebral hemisphericity as differential sources for the sensed presence, mystical experiences, and religious conversions. *Perceptual and Motor Skills* 76: 915–30.

Piaget, J. 1963. *The origins of intelligence in children.* New York: Norton. (Originally published 1952.)

Pinar, W., W. Reynolds, P. Slattery, & P. Taubman. 1995. *Understanding curriculum.* New York: Peter Lang.

Pinker, S. 1997. *How the mind works.* New York: Norton.

———. 2002. *The blank slate: The modern denial of human nature.* New York: Viking.

Plomin, R., M. Owen, & P. McGuffin. 1994. The genetic basis of complex human behaviors. *Science* 264: 1733–39.

Raymo, C. 1998. *Skeptics and true believers: The exhilarating connection between science and religion.* New York: Walker.

Ridley, M. 1996. *The origins of virtue: Human instincts and the evolution of cooperation.* London: Penguin.

Rosenau, P. 1992. *Post-modernism and the social sciences.* Princeton, NJ: Princeton University Press.

Ruse, M. 1986. *Taking Darwin seriously: A naturalistic approach to philosophy.* Cambridge, MA: Blackwell.

Russell, B. 2001. *The scientific outlook.* New York: Routledge. (Originally published 1931.)

———. 1975. *Religion and science.* New York: Oxford University. (Originally published 1935.)

———. 1957. *Why I am not a Christian and other essays on religion and related subjects.* New York: Simon & Schuster.

Sagan, C. 1996. *The demon-haunted world: Science as a candle in the dark.* New York: Ballantine.

Scharmann, L. 1993. Teaching evolution: Designing successful instruction. *American Biology Teacher* 55: 481–86.

Scott, E. 1997. Antievolution and creationism in the United States. *Annual Review of Anthropology* 26: 263–89.

Seckel, A., ed. 1986. *Bertrand Russell on God and religion.* Buffalo, NY: Prometheus.

Settlage, J. 1994. Conceptions of natural selection: A snapshot of the sense-making process. *Journal of Research in Science Teaching* 31: 449–57.

Shermer, M. 2000. *How we believe: The search for God in an age of science.* New York: W. H. Freeman.

Siegel, H. 1988. *Educating reason: Rationality, critical thinking, and education.* New York: Routledge.

Slezak, P. 1994. Sociology of scientific knowledge and scientific education: Part I. *Science & Education* 3: 265–94.

———. 1994. Sociology of scientific knowledge and scientific education: Part II: Laboratory life under the microscope. *Science & Education* 3: 329–55.

Smith, J. 1998. *Evolutionary genetics.* Oxford: Oxford University Press.

Sobel, D. 2000. *Galileo's daughter: A historical memoir of science, faith, and love.* New York: Penguin.

Sober, E., & E. Wilson. 1998. *Unto others: The evolution and psychology of unselfish behavior.* Cambridge, MA: Harvard University Press.

Sokol, A., & J. Bricmont. 1998. *Fashionable nonsense: Postmodern intellectuals' abuse of science.* New York: Picador.

Stachel, J., ed. 1998. *Einstein's miraculous year.* Princeton, NJ: Princeton University Press.

Sternberg, R., & T. Ben-Zeev. 2001. *Complex cognition: The psychology of human thought.* New York: Oxford University Press.

Tomasello, M., & J. Call. 1997. *Primate cognition.* New York: Oxford University Press.

Torrey, E. 1992. *Freudian fraud: The malignant effect of Freud's theory on American thought and culture.* New York: Harper Collins.

Trigg, R. 1986. Evolutionary ethics. *Biology and Philosophy* 1: 325–35.

Wandersee, J., J. Mintzes, & J. Novak. 1994. Research on alternative conceptions in science. In *Handbook of research in science teaching and learning,* ed. D. Gabel. New York: Macmillan.

Webster, R. 1995. *Why Freud was wrong: Sin, science, and psychoanalysis.* New York: Basic Books.

Weiner, J. 2000. *Time, love, memory: A great biologist and his quest for the origins of behavior.* New York: Vintage.

Whitehouse, H. 2000. *Arguments and icons: Divergent modes of religiosity.* Oxford: Oxford University Press.

Whitfield, P. 1993. *From so simple a beginning: An illustrated exploration of the 4-billion-year development of life on Earth.* New York: Macmillan.

Wilson, E. 1978. *On human nature.* Cambridge, MA: Harvard University Press.

———. 1998. *Consilience: The unity of knowledge.* New York: Knopf.

Wolpert, L. 1992. *The unnatural nature of science: Why science does not make common sense.* Cambridge, MA: Harvard University Press.

Wright, R. 1994. *The moral animal: Evolutionary psychology and everyday life.* New York: Random House.

Wright, W. 1999. *Born that way: Genes, behavior, personality.* New York: Routledge.

THREE PAPERS RELATED TO THE THEME OF THIS BOOK

Each of the three papers that follow focuses on the relationship between scientific and religious habits of mind. The first paper was published recently in the September 2003 issue of the *American Biology Teacher* and the others were presented at professional meetings of science educators. The author welcomes comments on the ideas in these papers and on any of the other material in this book. Send comments to: rgood@lsu.edu.

PAPER ONE:
EVOLUTION AND CREATIONISM: ONE LONG ARGUMENT

This paper was presented at the November 7–11, 2001, International Conference on History and Philosophy of Science in Science Teaching, Denver, Colorado. It describes an attempt in April 2001 by Louisiana lawmakers to pass a resolution declaring the writings of Charles Darwin racist, thereby impeding the teaching of scientific evolution in the public schools of Louisiana. This action by Louisiana lawmakers is typical of local and state actions across the United States to obstruct the teaching of evolution and other scientific knowledge deemed objectionable by creationists and others who feel their children are threatened by ideas inconsistent with creationism.

A Brief History

Ever since Darwin published *On the Origin of Species* in 1859, creationists have argued against the scientific theory of evolution. Creationists do not want God's hand removed from the creation of species and they see Darwin's dangerous idea (Dennett, 1995) doing just that. In *The Creationists*, Numbers (1993) provides an excellent history of the creationist movement, with emphasis on the so-called *scientific creationists* in the United States. A more recent book by Alters and Alters (2001) also examines the creation-evolution controversy, with more emphasis on ideas for biology teachers. Many other books, including those by Dawkins (1986, 1996), Eldridge (2000), Futuyma (1995), Gould (1989, 1999), Kitcher (1982), National Academy of Sciences (1998), Moore (2000), Numbers (1993), Pennock (1999), and Ruse (1982, 1989), document the evolution-creation controversy. The links between racism and the anti-evolution campaign have been documented recently by Moore (2000).

Related to the evolution-creationism struggle in Louisiana, the *Edwards v. Aguillard* (1987) decision of the U.S. Supreme Court held unconstitutional the Creationism Act of the Louisiana legislature. This law prohibited the teaching of evolution in public schools except when it was accompanied by instruction in "creation science." The law would have required the development of creationist teaching materials, curriculum committees, and related educational support. A decade after this decision, Don Aguillard, the biology teacher named in the *Edwards v. Aguillard* case, researched the factors influencing the teaching of biological evolution in Louisiana public schools as part of his doctoral degree requirements at Louisiana State University in Baton Rouge. As his major professor I encouraged Don to follow his earlier activities as a science teacher and central figure in the legal case with a careful analysis of the state of evolution education in Louisiana. Among his findings are the following (Aguillard, 1998):

- Forty-one percent of Louisiana public school biology teachers indicated either that creationism has a scientific foundation (24 percent) or that they were not sure (17 percent).
- There is a statistically significant correlation between instructional time devoted to evolution and beliefs regarding the validity of creationism.
- More than 75 percent of Louisiana public school biology teachers judged their academic training in evolution as inadequate.
- Most biology teachers report spending fewer than 5 hours (of about 180) dealing with evolution throughout the school year.

A decade after the *Edwards v. Aguillard* U.S. Supreme Court decision, the U.S. District Court for the Eastern District of Louisiana rejected a school board policy requiring teachers to read to students a disclaimer saying that evolution is *only a theory* (*Freiler v. Tangipahoa Parish Board of Education*, 1997). The decision identified *intelligent design* as equivalent to *creation science*, both promoted by religious fundamentalists who regard themselves as *creation scientists*. Further legal appeals by the Tangipahoa school board were unsuccessful.

Louisiana, like many other states in the United States, has tried repeatedly to suppress the teaching of evolution. The latest effort is described in the following section.

Darwin Ideology Is Racist

On April 15, 2001, in Baton Rouge, *The Advocate* announced that Louisiana state representative Sharon Weston Broome wanted lawmakers to pass a resolution rejecting "the core concepts of Darwinist ideology." According to Broome, evolution promotes racism because Hitler and others have used Darwin's writings to justify mass murder and other heinous crimes. Despite efforts by supporters of evolution education, the Louisiana House Education Committee passed the following resolution by a vote of nine to five.

Louisiana House Education Committee Resolution on Teaching Evolution

Whereas, America's fundamental document, "The Declaration of Independence," expresses the self-evident truth that all men are created equal, that they are endowed by their Creator with certain unalienable Rights, and that among these are Life, Liberty, and the pursuit of Happiness; and

Whereas, the Constitution of the State of Louisiana of 1974 declares that the only legitimate ends are to secure justice for all, preserve peace, protect the rights, and promote the happiness and general welfare of the people; and

Whereas, empirical evidence has documented an indisputable commonality among all people groups, or races, and has demonstrated that normal variations in the human gene pool account for our differences, of which racial differences are a trivial portion; and

Whereas, the writings of Charles Darwin, the father of evolution, promoted the justification of racism, and his books "On the Origin of Species by Means of Natural Selection: or the Preservation of Favoured Races in the Struggle for Life" and "The Descent of Man" postulate a hierarchy of superior and inferior races; and

Whereas, Adolf Hitler and others have exploited the racist views of Darwin and those he influenced, such as German zoologist Ernst Haekel, to justify the annihilation of millions of purportedly racially inferior individuals.

Therefore, Be It Resolved that the Legislature of Louisiana does hereby deplore all instances and ideologies of racism, does hereby reject the core concepts of Darwinist ideology that certain races and classes of humans are inherently superior to others, and does hereby condemn the extent to which these philosophies have been used to justify and approve racist practices.

Be It Further Resolved that the Legislature does also hereby urge and request the public education system of Louisiana, as appropriate in the curriculum, to address the commonalities of people groups and the weaknesses of Darwinian racism.

Be It Further Resolved that a suitable copy of this Resolution be transmitted to the commissioner of administration, who will make its contents known to the heads of each Louisiana state department and agency; to the Board of Regents, who will make its contents known to each college and university president or chancellor; and to the state superintendent of education, who will make its contents known to the superintendents of each city, parish, or other local public school system.

Following the action of the Louisiana House Education Committee, I developed a resolution on teaching evolution that was meant to represent the scientific community at Louisiana State University (LSU). The resolution was reviewed by several of my colleagues in various science departments at LSU and by other supporters of evolution education, and various suggestions were offered to strengthen the resolution. The resolution was then circulated among some of the faculty in the scientific community and signed by about one hundred people. Some of those who chose not to sign seemed to feel it would be better to remain silent rather than to confront the same legislature that rules on LSU's budget. Although one hundred may seem like a small number, it represents most of the faculty who actually had an opportunity to read and sign the resolution. The resolution in its final form is as follows:

LSU Resolution on Teaching Evolution

Whereas, the U.S. National Academy of Sciences, the National Science Teachers Association, the American Association for the Advancement of Science, and the National Association of Biology Teachers have continuously and strongly supported the teaching of modern evolutionary theory in our schools; and

Whereas, the U.S. Supreme Court and lower courts have ruled repeatedly that creationism in its various guises is religion, not science; and

Whereas, LA House Concurrent Resolution 74 is recognizable as an effort to discredit the teaching of evolution in Louisiana public schools and universities by incorrectly linking it to racism; and

Whereas, the LSU scientific community encourages the best possible science education opportunities for Louisiana students; and

Whereas, teaching biology without evolutionary theory is comparable to teaching physical science without atomic theory; and

Whereas, Charles Darwin stated in *Voyage of the Beagle* that "It makes one's blood boil, yet heart tremble, to think we Englishmen and our American descendents, with their boastful cry of liberty, have been and are so guilty" [of being involved in the slave trade]; and

Whereas, evolutionary biology more than any other area of science has undermined racism by showing the universal biological kinship of *all* humans; and

Whereas, the misuse of scientific knowledge has nothing to do with its validity;

Therefore, be it resolved that we members of the LSU Scientific Community are firmly committed to the teaching of evolutionary theory, atomic theory, and other scientifically valid knowledge in Louisiana's public schools.

Be it further resolved that we members of the LSU Scientific Community are firmly committed to the elimination of racism, sexism, and other forms of bigotry in our society by educating our citizens to understand that tolerance not censorship of science is the preferred path to enlightenment.

The signed copies of the resolution were collected and preparations were made to present them to key members of the Louisiana House of Representatives as they considered the Broome resolution. However, the Broome resolution was changed on the floor of the House, deleting all references to Darwin and Darwinist ideology, so the final version became simply a resolution to eliminate racism and other forms of bigotry in Louisiana public education. Even in Louisiana few legislators will publicly oppose such a resolution.

Attacking evolution by calling Darwin a racist is just one of many tactics used by those opposed to teaching evolution in our schools. At the local, state, and national levels religious fundamentalists and others opposed to evolution education continue to try to suppress the teaching of evolution and support school prayer and the teaching of their vision of religion in our schools. It is not just in the South or in Kansas that these battles are fought (Alters & Alters, 2001; Moore, 2000); all over the United States fundamentalists target evolution as the main enemy of their visions of God and religion. The long argument continues between evolutionists and creationists.

Taking History and Philosophy of Science Seriously

One would think that 140-plus years after Darwin published *On the Origin of Species* the most scientifically advanced nation on Earth would understand and embrace his ideas as important to a sound science education. However, science education research (Alters & Alters, 2001; Moore, 2000) and polls by Gallup (Gallup & Newport, 1991) and others (National Academy of Sciences, 1998; Sagan, 1996) reveal widespread ignorance among U.S. citizens regarding scientific evolutionary theory.

Both cognitive and religious obstacles to evolution education are real, persistent, and often interrelated. Trying to deal only with the cognitive obstacles (for example, misconceptions) while ignoring the religious/political obstacles seems likely to be less effective than taking both into account.

Fortunately, there are many good resources for biology teachers and others who want to help students understand the basics of evolutionary science and its implications. The eight-part PBS video series *Evolution* is one of the best instructional resources of its kind ever produced. In eight one-hour programs Darwin's ideas on evolution of life are presented, recognizing the potential controversy that might arise because of opposition by religious fundamentalists and others who dislike the implications of evolutionary science for human behavior. In addition to the eight one-hour programs, a free teacher's guide and an extensive web site (www.pbs.org) offer the interested teacher or parent or student valuable information on evolution and how to teach it.

The current position of the National Association of Biology Teachers (NABT), the National Science Teachers Association (NSTS), and similar professional groups is that religion and religious beliefs should be kept out of science classroom discussions of evolution and other scientific topics. This is also the position of our courts. The PBS program *Evolution* complicates this position because it recognizes that a major obstacle to evolution education is fundamentalist/creationist religious belief. From Darwin's day to the present there has been widespread opposition to evolution education in our public schools. *Taking history and philosophy of science seriously, where evolution is concerned, means that one must include adequate coverage of the role religion has played during the development of the theory of evolution.* The developers and producers of the PBS program *Evolution* realized this, beginning their eight-hour television series with a two-hour program entitled "Darwin's Dangerous Idea" and ending with a one-hour program entitled "What about God?" Historical facts about the development of Darwin's theory of evolution by natural selection automatically include religion and religious beliefs.

Two of the expert commentators in *Evolution* are Daniel Dennett and Stephen Gould, both well-known academics who have written widely about evolution and the nature of science. Dennett (1995) has taken the position that science and religion are deeply incompatible, but Gould (1999) disagrees, and the *Evolution* special seems to side with Gould. Although history suggests that Dennett is correct, the current public position among most scientific and science education organizations is that science and religion are separate spheres of activity with separate goals. In other words, science tells us about the heavens and religion tells us how to get to Heaven. When it comes to science education and religion a strong case can be made for Dennett's position (see Good, et al, 2001; Mahner & Bunge, 1996; and Martin, 1997, for supporting arguments), but the question here is—*Should history and philosophy of science be taken seriously where evolution education is concerned?* The answer is an unequivocal *Yes!* Just as Galileo's conflicts with the religious powers of his time should be a central part of physics education, Darwin's conflicts with religion should be a central part of biology/evolution education. This is what *Evolution* does and it is the educationally correct (although perhaps not politically correct) thing to do. Helping students see the personal struggles Darwin had with religion, including his own beliefs, humanizes the development of his theory of evolution and places it within a real social context. It may be the best and most important story science has to tell.

There are many other useful instructional resources for evolution education, including the National Academy of Sciences 1998 document *Teaching about Evolution and the Nature of Science*. It combines ideas about the nature of science with important ideas about evolution, and it includes many activities for students that are designed to help them better understand evolution. This guide and the PBS *Evolution* series are two very useful resources for the science teacher. A third instructional resource is the BSCS (2000) document *Genes, Environment, and Human Behavior*. This guide, by the best producer of evolution education materials since its beginning in 1958, introduces the reader to ideas and issues about the genetic components of human behavior. The authors explain the relationship between genetics and evolution:

Because genetics is the study of the root source of biological variation, which is central to evolutionary mechanisms, an understanding of basic principles in genetics is central to an understanding of evolution itself. (p. 11).

Religious fundamentalists' opposition to this area of study, behavioral genetics, is similar to their opposition to evolutionary science because they do not want to believe that humans are closely related to chimpanzees and other living things. And of course the "evils" of homosexuality are placed in a different light when the genetic components of human behavior are considered (Good, 2000; Hamer & Copeland, 1998).

The habits of mind associated with science are not only different from those associated with religion (and especially fundamentalist religion), they are basically incompatible. Most religions encourage believers to accept without evidence the authority of holy books and leaders, while science encourages a respect for real evidence and a questioning attitude toward authority. Nature is the final authority in science. When young children are indoctrinated into believing in that for which there is no evidence (God, Heaven, Hell, and so on) a habit of mind is being developed that is inconsistent with the open, inquiring mind needed for scientific study. The habits of mind are not merely different, they are incompatible between science and religion, especially where an unseen God or angels or other agents are said to intervene in people's lives. The fact that some scientists believe in a God that intervenes in our world, causing things to happen that otherwise would not, in no way *proves* that science and religion are compatible. It simply shows that some people are able to separate their lives as scientists from their religious lives. Of course Einstein's God (Jammer, 1999) and similar conceptions of God are not incompatible with a scientific outlook, but few people develop this kind of God belief. Most believe in a personal God who listens to prayers and occasionally answers them in some way.

The long argument between science and religion will continue because the habits of mind promoted by the two domains are basically incompatible. The PBS television program *Evolution* got it right. The story of evolution must include the history of the struggle between religion and science, not just the facts of evolutionary science. Understanding how scientific ideas are developed is just as important in science education as understanding the idea itself. One kind of knowledge without the other is insufficient to attain the goal of real scientific literacy.

REFERENCES

Aguillard, D. 1998. *An analysis of factors influencing the teaching of biological evolution in Louisiana.* Doctoral dissertation, Louisiana State University, Baton Rouge.

Alters, B., & S. Alters. 2001. *Defending evolution: A guide to the creation/evolution controversy.* Sudbury, MA: Jones and Bartlett.

BSCS. 2000. *Genes, environment, and human behavior.* Colorado Springs, CO: Author.

Dawkins, R. 1986. *The blind watchmaker: Why the evidence of evolution reveals a universe without design.* New York: Norton.

———. 1996. *Climbing Mount Improbable.* New York: Norton.

Dennett, D. 1995. *Darwin's dangerous idea.* New York: Simon & Schuster.

Eldridge, N. 2000. *The triumph of evolution and the failure of creationism.* New York: W. H. Freeman.

Futuyma, D. 1995. *Science on trial: The case for evolution.* New York: Sinauer.

Gallup, G., Jr., & F. Newport. 1991. Almost half of Americans believe Biblical view of creation. November, 30–34.

Good, R. 2000. Habits of mind associated with science and religion: Implications for science education. Paper presented at the Sixth International History, Philosophy, & Science Teaching Conference, November 7–11, Denver, CO.

Good, R., M. Hafner, & P. Peebles. 2001. Scientific understanding of sexual orientation: Implications for education. *American Biology Teacher* 62: 326–30.

Gould, S. 1989. *Wonderful life: The Burgess shale and the nature of history.* New York: Norton.

———. 1999. *Rocks of ages: Science and religion in the fullness of life.* New York: Ballantine.

Hamer, D., & P. Copeland. 1998. *Living with our genes.* New York: Anchor.

Jammer, M. 1999. *Einstein and religion.* Princeton, NJ: Princeton University Press.

Kitcher, P. 1982. *Abusing science: The case against creationism.* Cambridge, MA: MIT Press.

Mahner, M., & M. Bunge. 1996. Is religious education compatible with science education? *Science & Education* 5: 101–23.

Martin, M. 1997. Is Christian education compatible with science education? *Science & Education* 6: 239–49.

Moore, R. 2000. *In the light of evolution: Science education on trial.* Reston, VA: National Association of Biology Teachers.

National Academy of Sciences. 1998. *Teaching about evolution and the nature of science.* Washington, DC: National Academy Press.

Numbers, R. 1993. *The creationists: The evolution of scientific creationism.* Berkeley: University of California Press.

Pennock, R. 1999. *Tower of Babel: The evidence against the new creationism.* Cambridge, MA: MIT Press.

Ruse, M. 1982. *Darwinism defended: A guide to the evolution controversies.* Reading, MA: Addison-Wesley.

———. 1989. *The Darwinian paradigm: Essays on its history, philosophy, and religious implications.* New York: Routledge.

Sagan, C. 1996. *The demon-haunted world: Science as a candle in the dark.* New York: Ballantine.

PAPER TWO: USING A SCIENCE LITERACY PRETEST IN A SCIENCE TEACHER EDUCATION COURSE

This paper reports on research done by this author as instructor in a course for preservice science teachers at Louisiana State University in Baton Rouge. The results of the study point to a serious problem with the level of understanding of scientific knowledge, especially evolution and the nature of science, by science teachers.

The Course

"Critical Issues in Science Teaching" is a course for preservice science teachers at LSU usually taken the semester before a science methods course. Among the issues considered in the course are (1) teaching about evolution and the nature of science, (2) student misconceptions in the physical sciences, (3) using history and philosophy of science in science teaching, (4) national and state science educational standards, (5) assessment, (6) using the Internet as a resource, and (7) human population growth and its implications. The emphasis is on issues rather than methods of teaching, but there is not a sharp line drawn here so it is not uncommon to discuss and sometimes demonstrate teaching strategies. Also, the students are required to observe and collect data on science teaching in local schools.

Among the texts and related resources used in the course are: *Teaching about Evolution and the Nature of Science* (NAS, 1998), *Targeting Students' Science Misconceptions* (Stepans, 1996), *Science Teaching: The Role of History and Philosophy of Science* (Matthews, 1994), *National Science Education Standards* (NRC, 1996), and various videos and web sites. For the human population growth section of the course, a Zero Population Growth (ZPG) workshop leader visits the class to show their famous world population simulation video and to do various activities. Other topics are included, but this should give the reader a good idea of what is meant by critical issues. Any of the issues could be included in a science methods course, but the emphasis here is on *reasons* for doing various things in a science class rather than the methods that might be used.

The Pretest

The pretest is shown here just as it is given to the students on the first class meeting.

Science Concepts Questionnaire
    Name_____
    Date_____
    Science Major_____
    Science Minor_____
1. Approximate age of the Earth:
2. Approximate age of life on Earth:
3. Approximate age of Homo sapiens:
4. Do you believe humans are closely related to chimpanzees?
   Explain:
5. How does the mechanism of *natural selection* work?
6. Do you agree with the following statement: "If evolution is taught in public schools then creationism should be taught too."
   Explain:
7. Do you agree with the following statement: "The fossil record is so full of gaps that one cannot have confidence that evolution really occurred."
   Explain:
8. Do you agree with the following statement: "Evolution is just a theory and other theories should be considered as well in teaching biology."
   Explain:
9. Do you agree with the following statement: "We can probably learn a lot about humans by studying other animals."
   Explain:
10. Do you think it is likely that angels exist?
    Explain:
11. Do you think it is likely that ghosts exist?
    Explain:
12. How do scientists explain the occurrence of seasons (summer, fall, winter, spring) on Earth?
13. How far in miles is our moon from us?
14. How far in miles is our sun from us?
15. How far is the nearest star from us?
16. Why do astronauts experience weightlessness as they circle the Earth?
17. What causes the phases of our moon?
18. What factor(s) determine the period (swing time) of a pendulum?
19. Why did the Catholic church in Rome accuse Galileo of heresy?
20. What is the meaning of Einstein's equation $E = mc^2$?
21. About how fast, in miles per hour, does sound travel?
22. About how fast, in miles per hour, does light travel?
23. What is the distance, in miles, around Earth's equator?
24. How do scientific habits of mind compare to religious habits of mind?
25. Describe a scientifically literate person.

The first nine items are designed to assess ideas/beliefs related to evolution of life, the main topic of the text, *Teaching about Evolution and the Nature of Science*, and other items are designed to assess ideas/beliefs related to the physical sciences, nature of science, and religious beliefs. Items 12 and 17 are the focus of the well-known video *A Private Universe*. Most of the items seem to fall within precollege levels of science literacy as defined in *Benchmarks for Science Literacy* and *National Science Education Standards*.

The Students

The 25 students in this particular section of the "Critical Issues" course included 16 biology education majors, 4 chemistry education majors, and 5 students who were in other categories (for example, in-service teachers, an aviation science major). Most of the 25 students were seniors (13) and juniors (8).

Results and Discussion

Pretest data on the 16 biology education majors and the 4 chemistry education majors will be presented and discussed, omitting the *other* category.

Ideas/Beliefs Related to Evolution

Two of the biology education majors and 1 of the chemistry education majors responded to items 1 through 3 by writing 6,000 years after each one. So 3 of the 20 students who want to teach science to our children seem to be *young-Earth creationists*. Their responses to questions 4 through 11 confirm this suspicion. You might think that having only 3 creationists out of 20 is not too bad, but it gets worse. Of the remaining 17 students, 5 (4 biology and 1 chemistry) said yes to items 7 and 8, so they seem to disagree that evolution is a well-established scientific theory. So 8 of the students are either young-Earth creationists or they do not have confidence in evolution as a scientific theory. Thus far it appears that 40 percent of these future science teachers are not prepared to do what the authors of *Teaching about Evolution and the Nature of Science* recommend; that is, use evolution as the organizing theme in biology.

How do the remaining 60 percent of the prospective science teachers (12 biology and 2 chemistry) respond to the other items designed to assess ideas/beliefs about evolution? Four (3 biology and 1 chemistry) are not too sure about item 7 (that is, they have low confidence in evolution) and they agree that other "theories" (such as creationism) should be taught in our public schools. The score is now down to the 40 percent mark, so less than half of the prospective science teachers remain, and there are questions about some of them (7 biology and 1 chemistry). However, rather than further analyze their responses to items 1 through 9, let's assume that the 7 prospective biology teachers are (or will be soon) in a reasonably good position to begin their careers as biology teachers, regarding their understanding of evolution as the central organizing theme in biology. This means that fewer than half of the biology education majors can be considered scientifically literate regarding evolution, even though all have taken at least 24 semester hours of college biology courses. What is the likelihood that the majority will change their ideas/beliefs about evolution by the time they are first-year teachers? Furthermore, how likely is it that the 7 prospective biology teachers who seem to understand the proper relationship between science (evolution) and religious belief (creationism) will actually treat evolution as the central organizing theme in biology, especially if there is opposition from parents and others? The author leaves it to the reader to consider this question.

Ideas/Beliefs Related to Physics/Astronomy

Eleven of the pretest items (12–18, 20–23) involve knowledge that often comes under the heading of physics. These items are meant to relate to a second major issue in science teaching, *misconceptions* held by students and related conceptual change theory. It is generally agreed that once students become aware of their own pre-scientific conceptions, they are in a better position to make progress toward a more scientifically accurate conception.

Like the graduating seniors in the video *A Private Universe*, these students do not seem to have scientific explanations for the occurrence of seasons on Earth (item 12) or for the phases of our moon (item 17). For item 12 only 7 of the 20 students mentioned the tilt of the Earth in relation to the sun's incoming light rays and of these 7 only 3 seemed to understand what it was about the tilt that actually caused the seasons. None of the 20 students gave what could be considered a scientifically correct reason for the phases of our moon (item 17). It should be mentioned here that LSU requires high school physics as an admission requirement so it is safe to assume that most of these 20 students passed high school physics. Also, given that most upper elementary and middle school science textbooks cover these topics these results are even more discouraging.

For item 16, why astronauts experience weightlessness as they circle Earth, 19 of 20 students said it was because they were too far from Earth's gravity. When I pointed out to them that the astronauts were only about 200 miles above Earth's surface (that is, about 4,200 miles rather than 4,000 miles from Earth's center), they were completely stumped. One student said it was centripetal force but she was unable to explain what that meant.

Most students had no idea how far away the sun and moon are from Earth. One student said 5,000 miles for the moon and 1.5 billion miles for the sun and another student said 1,000 and 1 billion miles respectively. And for the distance in miles around Earth's equator, 7 students had no idea, 6 students were within 10,000 miles of 25,000, and the remaining 7 estimates were from 3,645 miles to 5 million miles.

Overall it can be said that these 16 biology education majors and 4 chemistry education majors did not do any better on the physical science items than they did on the evolution items. In fact, they did worse.

Other Test Items

The other items ask about the likelihood of existence of angels (item 10) and ghosts (11), how scientific and religious habits of mind compare (24), and a description of a scientifically literate person. Most believe in the existence of angels but they are less sure of the existence of ghosts, although the reasons for the difference are not clear. Answers to question 24 suggest that most understand that science requires natural evidence for its theories while religion does not, relying instead on faith in the accuracy of holy books or authorities. Most of the students saw no real conflict between religion and science even though they were aware of historical conflicts such as Galileo and the Catholic Church, and Darwin and religious leaders of his time. Among the few who saw a conflict between scientific and religious habits of mind is a biology education major who said this: "Both are such serious ideologies that their followers are willing to doubt the other field even exists. One difference is scientists are constantly trying to disprove their theories while religious fanatics are constantly trying to prove their God exists."

In trying to describe a scientifically literate person most said it was someone who understands science concepts and how science is done. By this definition and based on the results of the pretest, it would be difficult to characterize many of these students as scientifically literate.

Discussion

So what does all this mean? How can such a pretest be helpful in preservice science teacher education courses? First, it alerts the instructor to some of the deficiencies in students' basic science knowledge. In a class of mostly biology education majors, it is clear that most have a shaky understanding of the nature of evolution as the central organizing theme in biology. And even more troubling, it emphasizes the likely existence of anti-evolution beliefs among preservice biology teachers. It explains, in part at least, why we have so many practicing high school biology teachers who do so little to help their students understand the central role of evolutionary theory in biology (Aguillard, 1998).

Second, such a pretest helps students focus on the nature of their own science knowledge. Most of the students in this sample thought the test was very difficult but not unfair, although there was clearly some discomfort with the evolution and creationism issues. They were very interested in how their class peers responded to the various items and they posed some of the questions to their friends and family members. The test was a good introduction to two of the texts (*Teaching about Evolution and the Nature of Science* and *Targeting Science Misconceptions*) used in the course.

Third, using such a pretest demonstrates the need to use a similar strategy with secondary school students. Asking students to make predictions and provide explanations for natural events seems to heighten the interest of students in a classroom setting. Starting each science class with a question about some aspect of nature and trying to figure out how various claims or hypotheses might be tested seems to get at the heart of inquiry teaching and learning as suggested in *Benchmarks* and *Standards*.

It takes time to administer and score the kind of pretest described in this paper, but it can be an eye-opening experience for both the instructor and the students. A science teacher education issues or methods course should not avoid dealing with science content, as it is (or should be) the central issue in science education. The science teacher must be proficient in the science content as well as the instructional strategies used to help students better understand that content. Emphasizing inquiry in science teaching and learning does not excuse the teacher from understanding the science she is about to teach.

Administering a pretest on the first day of a science education course is not the only way to emphasize the importance of understanding science content before you attempt to teach it, but it really gets the attention of the preservice science teachers. They have to come to grips with what it means to understand something before you are in a good position to teach it to others. It is clear that taking a few college courses in biology does not ensure that a student will understand the central role of evolution, for example, or that creationism is belief based on faith, not scientific evidence, and, therefore, cannot contribute to a scientific understanding of nature. Science teaching methods or issues courses should stay close to science content and the nature of science as various teaching methods or issues are considered, and one way of doing that is to use the kind of pretest described in this paper.

REFERENCES

Aguillard, D. 1998. *An analysis of factors influencing the teaching of biological evolution in Louisiana public secondary schools.* Unpublished dissertation, Louisiana State University, Baton Rouge.

Matthews, M. 1994. *Science teaching: The role of history and philosophy of science.* New York: Routledge.

National Academy of Sciences. 1998. *Teaching about evolution and the nature of science.* Washington, DC: Author.

National Research Council. 1996. *National science education standards.* Washington, DC: National Academy Press.

Stepans, J. 1996. *Targeting students' science misconceptions: Physical science concepts using the conceptual change model.* Riverview, FL: Idea Factory.

PAPER THREE: WILL HUMAN BEHAVIORAL GENETICS/SEXUAL ORIENTATION BE THE NEXT TARGET OF THE CENSORS OF SCHOOL SCIENCE?

This paper was presented at the AETS International Meeting in Costa Mesa, California, January 18–21, 2000, as part of the symposium *Preparing Teachers to Address Anti-Science Censors: The Cases of Evolution and Behavioral Genetics*. In this paper I provide a brief history of human behavioral genetics, summarize a related study of four high school biology classes, and discuss ways to resist censorship attempts.

A Brief History of Human Behavioral Genetics

Charles Darwin's Work

Darwin was careful to avoid discussion of human origins and behavior in *The Origin of Species*, fearing that it would add to prejudices against his views on evolution by natural selection. However, in the years following 1859 he decided to provide details to his comment in *The Origin* that "light would be thrown on the origin of man," and in 1871 his *Descent of Man* appeared, followed by *The Expression of Emotions in Man and Animals* in 1872. He credits others, including Wallace, Huxley, and Haeckel, with recognizing that humans descended from lower forms. However, because of the comprehensiveness of his books of 1871 and 1872, and his earlier *Origin*, many historians and scientists trace human behavioral genetics to Charles Darwin. He begins the first chapter of *Descent* ("The Evidence of the Descent of Man from Some Lower Form") with the following statement: "He who wishes to decide whether man is the modified descent of some pre-existing form, would probably first enquire whether man varies, however slightly, in bodily structure and in mental faculties." This is followed by a great deal of evidence that makes a convincing case for human behavioral genetics. Finally, in chapter 21 ("General Summary and Conclusions") Darwin concludes with these words:

> The main conclusion arrived at in this work, namely, that man is descended from some lowly organized form, will, I regret to think, be highly distasteful to many.... We must, however, acknowledge, as it seems to me, that man with all his noble qualities, with sympathy which feels for the most debased, with benevolence which extends not only to other men but to the humblest creature, with his god-like intellect which has penetrated into the movements and constitution of the solar system—with all these exalted powers—Man still bears in his bodily frame the indelible stamp of his lowly origin.

A little more than a century later, with the appearance of Edward Wilson's *Sociobiology* in 1975, the linkage of human behavior to other animals continues to attract controversy.

Gregor Mendel's Work

When Gregor Mendel's paper *Experiments in Plant-Hybridization* was published in 1865 Darwin was unaware of the work and apparently never read or learned of the ideas about inheritance. And when Mendel's work was rediscovered around 1900, many biologists saw inconsistencies between the particulate nature of inheritance and evolutionary theory. They did not see how the continuous variation of evolution by natural selection could be explained in terms of the discontinuous variants displayed by Mendel's ratios, clearly representing particulate, discontinuous phenomena. These problems were gradually solved during the 1920s and '30s as evidence was found for continuous traits being caused by multiple genes that have small, additive effects. The *grand synthesis* of genetics and evolution occurred during the 1930s and '40s as population genetics grew and biologists resolved earlier inconsistencies between the two disciplines. As noted in the BSCS publication *Genes, Environment, and Human Behavior:*

> Because genetics is the study of the root source of biological variation, which is central to evolutionary mechanisms, an understanding of basic principles in genetics is central to an understanding of evolution itself. (BSCS, 2000, p. 11).

Francis Galton and Eugenics

Francis Galton (1822–1911) is usually credited (or blamed) with starting the field of behavioral genetics and he invented the term *eugenics* to mean well born. Strongly influenced by Darwin's *Origin of Species*, Galton collected data on human family resemblances, including intelligence, and concluded that many behavioral traits were inherited. Galton apparently ignored the influence of environment, and his writings were very influential within academic and political circles. During the early part of the twentieth century eugenics in the United States gained momentum and many states passed laws in the 1920s and '30s for compulsory sterilization of mentally retarded citizens. The Eugenics Sterilization Law passed in Germany in 1933 was influential in setting the stage for the Holocaust. The beginning of the human behavioral genetics movement was a disaster and the eugenics movement continues to be used as a club to criticize any linkage between biology and human behavior. The *environmentalists* consider nurture to be the only viable option in the nature-nurture debates and during most of the 1940s through the 1980s they dominated both the academic world and the political world. However, a few scientists were willing to consider the possibility that both nature and nurture influence human behavior.

Edward Wilson and Sociobiology

When *Sociobiology: The New Synthesis* appeared in 1975, its author, Ed Wilson, noted biologist at Harvard University, argued in his last chapter that humans too are instinctively social animals. The environmentalists, devoted to the nurture part of the nature-nurture debate, vilified him for stepping over the line into human behavior. It was one thing to speak of biologically induced social behavior among ants or bees or even chimpanzees, but quite another to suggest that human behavior also is influenced by our biology. Although Wilson's *Sociobiology* was eventually rated by the Animal Behavior Society as the most important book on animal behavior of all time, the decade following its publication was filled with opposition to Wilson and his ideas on human nature. The wave of opposition from social scientists committed to nurture at all costs, and from some biologists, especially Richard Lewontin and Stephen Gould at Harvard, was intense. In his excellent autobiography, *Naturalist*, Wilson claims that he was surprised at the intensity of the hostility provoked by his book, especially from his two natural science colleagues at Harvard. According to Wilson, the attacks were primarily on political rather than evidentiary grounds, and Lewontin's attacks were the most vehement:

> By adopting a narrow agenda of publishable research, Lewontin freed himself to pursue a political agenda unencumbered by science. . . . He disputed the idea of reductionism in evolutionary biology, even though it was and is the virtually unchallenged linchpin of the natural sciences. And most particularly, he rejected it for human social behavior. (Wilson, 1994, p. 345).

Wilson and his colleagues who were committed to the evidence of gene-culture co-evolution continued to publish their work in books and articles, and most scientists now agree that human behavior is the product of complex interactions between nature and nurture. However, many social scientists, religious fundamentalists, and those still committed to setting humans apart from other animals refuse to consider the overwhelming evidence in favor of nature-nurture interactions. Unlike Darwin's theory of evolution by natural selection, which is opposed mainly by religious fundamentalists on the political right, human behavioral genetics is opposed by many academics on the political left as well.

Objections to Human Behavioral Genetics/ Sexual Orientation Research

Some of the objections to human behavioral genetics have been identified in the previous section. From both ends of the political spectrum, godless Marxists and religious fundamentalists object to human behavioral genetics or evolutionary psychology and they try, in various ways, to censor the voices of their opponents. The rightist fundamentalists want humans to have a soul that sets them apart from mere animals, and academic leftists want humans to be fully in charge of their behaviors. Of course, one need not be sympathetic religious fundamentalism or Marxism to oppose a role for biology in human behavior. There are many strange bedfellows in that camp in addition to the two extremes. The behaviorists want to avoid theories of mind that complicate their world, and there are many remnants of that camp who continue to ignore evidence coming from the neurosciences that dissolves the distinction between mind and brain. The loose collection of academics who think of themselves as postmodernists object to any explanation of human behavior (or just about anything else) that seems to place itself above other equally valid local, ethnic ideas. And many philosophers in academia want to explain all human decision-making in terms of rational thought in the tradition of ancient scholars such as Plato and Aristotle.

Looking to chimpanzees, other mammals, and even so-called lower animals for explanations of human nature seems to many people to be somehow inappropriate. But if evolution and modern DNA studies link us closely to the rest of the animal world, how can we not look in that direction for clues to human nature? This question has been asked by more and more people in recent years and is certain to be at the center of attention in coming years. As evidence from the various neurosciences, including behavioral genetics, continues to mount, it is nearly certain that all but the most closed-minded critics will begin to admit that nature does play an important role in human behavior.

Of the various human behaviors of interest to researchers and the general public, none is more interesting or controversial than sexual orientation. Although Freud was wrong about most things in his theoretical musings (Webster, 1995), he was right about sex being interesting. Sexual orientation is, or should be, of interest to educators because of the fact that so many adolescents and young adults of the minority sexual orientation commit suicide. McKee and colleagues (1993) estimate that gay and lesbian youths constitute 30 percent of all youth suicides. This means that they commit suicide at a rate six times higher than their heterosexual peers. All parents, educators, and others who say they care about children should be concerned that so many young people feel bad enough to end their lives. To the extent that our negative climate for homosexuals contributes to their decision to commit suicide, we must try to do something to change that climate. In the next section I describe some ways science teacher educators and others can try to change the negative climate created for homosexuals by heterosexuals in this country. This includes being prepared for attempts to censor valid science knowledge about the naturalness of homosexuality in humans and other animals.

A Role for Science Teacher Educators

The long fight against the teaching of Darwinian evolution waged by the creationists (see Numbers, 1993, for an in-depth account) continues and there is no good reason to believe it will cease any time soon. The best preparation for such attacks on science and rationality is to be well informed about the science and about the tactics used by the opposition. In the case of evolution the opposition consists mainly of creationists who interpret their holy books literally, but for human behavioral genetics the opposition includes a more diverse group. Understanding the motivations and tactics of those opposed to teaching valid science can be helpful, and one of the tactics used by those opposed to teaching human behavioral genetics is to wave the red flag of eugenics. Although it is certainly true that Galton's eugenics movement was based on misguided and sloppy science and resulted in many violations of human rights, including compulsory sterilization of certain U.S. citizens, to continue to use eugenics as a blockade to behavioral genetics is an inappropriate strategy.

Traditional Assumptions about Human Nature

Human behavioral genetics challenges many traditional assumptions in society. Among these assumptions are the following:

- People have free will and so can be anything they want.
- People are unrelated to other animals.
- A person's temperament is influenced mostly by the environment, especially parental behavior.

These assumptions are widespread and can cause people to draw many unfounded conclusions about human nature. It is not possible in this brief paper to provide a detailed critique of these assumptions, but I explain how they can be misleading and then suggest further reading for more details.

In our democratic society all people are presumed to be free to believe what they want and become whatever their motivations suggest, short of interfering with others' rights under the law. The free-will doctrine is a deeply held political belief that carries over into many aspects of our lives, including our ideas about human nature. The principle that all avenues of life should be open to all citizens influences us in ways that are difficult to measure. However, nature places general limits on us that can be very difficult to overcome, and physical limits like height are much easier to see than those hidden from sight in our brains. We can try to become anything we might desire, but reaching our goals by overcoming our biological tendencies is another story. Assumptions 1 and 3 are closely related in this regard.

Assumption 2, *People are unrelated to other animals*, is widespread and especially strong within many religions. Many, perhaps most, people do not think that the study of other animals could have much to say about human nature. Our ethics, for example, are seen as unique to humans and derived from religion rather than somehow related to other primate behavior. Recent DNA studies showing that we share about 98 percent of our DNA with chimpanzees, and primate behavior studies showing that many human emotions and behavior are found in primates should cause some to question this assumption, but we will have to wait and see what happens.

And finally assumption 3, *A person's temperament is influenced mostly by the environment, especially parental behavior*, is held by most people. To some extent the pseudoscience of Sigmund Freud continues to be influential, even though his ideas have been discredited by the scientific community for decades (see Webster, 1995, for a thorough account), and our common sense suggests that parenting should have a strong influence on our basic personality and temperament. The book *Living with Our Genes* (1998), by NIH scientist Dean Hamer and Peter Copeland, looks at research in genetics, molecular biology, and psychology and concludes:

> People are unique from the moment of conception, they do not begin as indistinguishable lumps of stone sculpted by life into individuals. Each of us is born into the world as someone; we spend the rest of our lives trying to find out who (p. 25).

The influence of these assumptions is strong and complex, affecting much of our outlook on who we are and how we got that way. The book by Hamer and Copeland (1998), the BSCS publication mentioned earlier, and related work by Burr (1996), Cloninger, et al. (1994), Eysenck (1967), Kagan (1994), Rowe (1994), Wright (1999), and Zuckerman (1994) provide a comprehensive look at the various influences of our biology on our temperament and behavior.

Recognizing and changing these assumptions can lead one to the current scientific understanding of human nature and behavior as expressed in the BSCS module *Genes, Environment, and Human Behavior* (2000):

> Genes, organisms, and environments interact so that each is both cause and effect in a complex but increasingly analyzable way (p. 7).

Students' Ideas and Attitudes

A recent study (Good et al., 2000) of students in four high school biology classes suggests that human behavioral genetics, and sexual orientation in particular, can be taught effectively to high school freshmen. During six hours of instruction over three days, the teacher provided information and led activities on sexual determination and sexual orientation. Relying heavily on information and activities from the BSCS module *Genes, Environment, and Human Behavior* and other instructional materials such as video segments from *The Brain* and *Secrets of Life* series, information from Hamer's book *Living with Our Genes*, and related sources, the teacher actively engaged the students during each double-period class. Data collected using a questionnaire and from the observations of the teacher and two observers (including this author) resulted in the following conclusions:

- *Before* instruction students have little knowledge of the biological/genetic basis of human behavior.
- *Before* instruction most students are not aware of the role of brain chemicals such as serotonin in influencing certain personality traits.
- *Before* instruction most students believe sexual orientation is mainly a choice.
- *Before* instruction most students believe homosexuality is wrong whether it originates in our genes or in our culture.
- *After* instruction students have more knowledge of the biological/genetic basis of human behavior.
- *After* instruction most students are aware of the role of brain chemicals in influencing certain personality traits.
- *After* instruction students are less sure that homosexuality is mainly a choice.
- *After* instruction students are slightly less sure that homosexuality is wrong whether it originates in our genes or in our culture.
- *After* instruction students understand that human DNA differs from chimpanzee DNA by only 2 percent, but they still disagree that we can probably learn a lot about sexual determination and orientation in humans by studying other animals.

This study, with an experienced teacher and mostly ninth-grade students in four high school biology classes, makes it clear that the topics of sexual determination and sexual orientation can be dealt with effectively. The classroom teacher and this author agree with the authors of the BSCS module *Genes, Environment, and Human Behavior* that human behavioral genetics should be included in the high school biology curriculum as a major topic.

Preparing Teachers to Resist Censorship Attempts

The completion of the decoding phase of the Human Genome Project and the publication of the BSCS module signal the need for more information on these and related topics in the precollege biology/life science curriculum. The ethical, social, and legal implications of the emerging sciences focused on humans are considered by the authors of the BSCS module and should become part of the school curriculum as well. It is likely that social studies teachers will become involved in discussions of the social implications of applying new scientific knowledge about people. Wilson (1998) argues that social scientists must work with natural scientists to build bridges between the natural and social sciences—and that is happening in fields such as human behavioral genetics and evolutionary psychology. A natural extension of these collaborations among scientists is to have more communication between precollege science and social studies teachers. As more of the biology curriculum is devoted to human behavioral genetics, including sexual orientation, teachers will need to be well prepared to teach the science content and, as with evolution, they will need to understand the various opponents/censors of these new ideas.

The best preparation is a sound knowledge of the science content and of various instructional tools that can be used effectively with students. Also, being aware of common objections raised by opponents of human behavioral genetics is important. Two of my colleagues and I (Good et al., 1999) cautioned science teachers to be prepared for censorship challenges and to be sensitive to the motivations behind those who may sincerely believe that students must be protected from ideas that threaten their belief systems. Whether the censorship threats come from anti-evolution creationists, postmodern critics of science, proponents of ethnic or multicultural sciences, or those opposed to biological/genetic interpretations of human behavior, it is important to understand the reasons behind their opposition. Especially for the last category of objections to human behavioral genetics, we must be aware of the very real threats to our freedom and privacy rights posed by improper use of genetic information. A great many thoughtful people have reservations about how genetic information might be used against certain citizens. Will insurance companies use genetic maps for this purpose? Will parents and their doctors try to design babies with certain traits? The potential for misuse of genetic information is great and will require much wisdom to develop and implement policies and laws that protect individuals and groups against the prejudices of the powerful.

When should genetic information be used to change the natural course of events? In the case of terrible disease nearly all agree that genetic manipulation should be allowed if it can improve things. It is not the extreme cases that are problematic but the cases involving traits that represent more modest disadvantages to attaining a happy, productive life. Parents want children to have the best chance for a happy, productive life, but how far should they be allowed to go in using available genetic technology to attain this goal? Who decides and how are decisions reached? These are issues that should be dealt with in our schools, and science teachers must be involved. The BSCS module *Genes, Environment, and Human Behavior* that has been mentioned throughout this paper recommends that biology teachers involve their students in discussions related to these issues. Those responsible for preparing science teachers must be sure to include these issues in teacher preparation programs and in-service workshops for practicing teachers. Work with those responsible for social studies teacher preparation and encourage more cooperation between science and social studies teachers in our schools. This is an area where more collaboration of this sort is needed and it needs to happen fast before others with less noble motives (money, for example) influence public policy in a direction that does not serve the interests of a humane, scientifically literate citizenry.

REFERENCES

BSCS. 2000. *Genes, environment, and human behavior.* Colorado Springs, CO: Author.

Burr, C. 1996. *A separate creation: The search for the biological origins of sexual orientation.* New York: Hyperion.

Cloninger, C., D. Svrakic, & T. Przybeck. 1994. A psychobiological model of temperament and character. *Archives of General Psychiatry* 50: 975–90.

Eysenck, H. 1967. *The biological basis of personality.* Springfield, IL: Charles C. Thomas.

Good, R., P. Peebles, & M. Loupe. 2000. Human behavioral genetics/sexual orientation in high school biology: Problems and prospects. Paper presented at the annual meeting of the National Association of Biology Teachers, Orlando, FL, October.

Good, R., J. Shymansky, & L. Yore. 1999. Censorship in science and science education. In *Caught off guard: Rethinking censorship and controversy*, ed. E. Brinkley. Boston, MA: Allyn & Bacon.

Hamer, D., & P. Copeland. 1998. *Living with our genes.* New York: Anchor.

Kagan, J. 1994. *Galen's prophecy: Temperament in human nature.* New York: Basic Books.

McKee, P., R. Jones, & R. Barbie. 1993. *Suicide and the school: A practical guide to suicide prevention.* Horsham, PA: LRP Publications.

Numbers, R. 1993. *The creationists: The evolution of scientific creationism.* Berkeley: University of California Press.

Rowe, D. 1994. *The limits of family influence: Genes, experience, and behavior.* New York: Guilford.

Webster, R. 1995. *Why Freud was wrong: Sin, science, and psychoanalysis.* New York: Basic Books.

Wilson, E. 1994. *Naturalist.* Washington, DC: Island Press.

———. 1975. *Sociobiology: The new synthesis.* Cambridge, MA: Belknap/Harvard University Press.

Wilson, E. 1998. *Consilience: The unity of knowledge.* New York: Knopf.

Wright, W. 1999. *Born that way: Genes, behavior, personality.* New York: Routledge.

Zuckerman, M. 1994. *Behavioral expressions and biosocial bases of sensation seeking.* Cambridge: Cambridge University Press.

INDEX

A
Aguillard, Donald, 79–80, 91
altruism, 32
argument from design, 23
Armstrong, Karen, 36

B
Bachelard, Gaston, 7
behaviorism, 31
Benchmarks for Science Literacy, 39
Bishop, George, 42
blank slate, 56–57, 60
Boyer, Pascal, 28, 33, 60
brand loyalty, 49
Bricmont, Jean, 53
Broome, Sharon, 80
BSCS, 84, 85, 94
Bunge, Mario, 46, 84

C
catastrophism, 8,
Catholic church, 1–3, 27, 36
censorship, 93, 99–100
cognitive dissonance reduction, 28, 33
common descent, 13
computational theory of mind, 31
confirmation bias, 33
consensus effect, 28, 33
consilience, 44, 50–52, 55–56
Copernicus, Nicholas, 1–3, 27
cosmic religious feeling, 18–19, 43
creation science, 20, 54, 79–80
creationism, 8, 12, 41, 47, 55, 64
Cummins, Catherine, 73

D
Darwin, Charles, 5-year voyage, 3, 8 13, 47–48;
on natural selection, 3, 8–11, 13–14, 15, 25–27, 50;
Origin of Species, 4, 13–14, 48, 51, 79, 81;
Descent of Man, 4, 15–16, 45, 48;
on religion, 5, 7, 9, 12, 14;
on evolution and humans, 5, 8, 11, 13–15, 18, 22, 34, 46;
and fundamentalists, 11, 25
Dawkins, Richard, 33, 37

demon-haunted world, 42, 63
Dennett, Daniel, 14, 31, 44–45, 79, 84
Democracy and Education, 39, 49–50
Dewey, John, 39, 43, 49–50
Dobzhansky, Theodosius, 5

E
Einstein, Albert, 3, 5–7, 17–25, 39, 43, 50, 62, 66, 85, 88
emotions, 4–5, 16, 32, 52, 60, 93, 97,
empiricist view, 45
essentialism, 13
eugenics movement, 56, 94, 96
Evolution: A Journey, 45
evolution education, 40, 44, 67, 80
evolution education research, 67

F
FAQs, 48–50
First Amendment, 41, 55, 64
free will, 61, 97
Freud, Sigmund, 58–59, 96–97
Freudianism, 31

G
Galileo, 1–3, 84, 88, 91
Galapagos Islands, 9–11
Galton, Francis, 94
God beliefs/experiences, 19, 28–29, 37, 45, 64
Goldsmith, Timothy, 48
Gould, Stephen, 10, 84, 95

H
Hamer, Dean, 98
History of God, 36–37
human behavioral genetics, 93–95
human nature 95–98

I
informed skepticism, 26
intelligent design, 80
irreducible complexity, 23

J
Jammer, Max, 18–20, 85

K
Kepler, Johannes, 1–3

L
Lewontin, Richard, 95
Loftus, Elizabeth, 59
Lyell, Charles, 8, 13, 25, 47

M
Mahner, 46, 84
Malthus, Thomas, 4, 8, 10–11, 13
Maxwell, James, 5
Mayr, Ernst, 12–14, 26, 44
meme, 33, 37
memory illusions, 28, 33
Mendel, Gregor, 94
mind as blank slate, 56
misconceptions, 45, 021–22
monotheism, 36
multicultural science, 54, 100

N
NABT, 47, 83
National Science Education Standards, 31, 40
National Science Foundation, 67
NAS, 87
natural selection, 3–7, 9, 11, 13–17, 23, 25–27, 31, 40–41, 45, 50, 57, 61, 81, 84, 88
natural theology, 13
naturalistic fallacy, 45–46
nature-nurture, 57, 94–95
Newton, Isaac, 1, 3
Newton, Roger, 26
Numbers, Ronald, 79

O
openness in science, 26

P
PBS/WGBH, 45
Persinger, Michael, 28–29
physicalism, 26
Pinker, Stephen, 3133, 56–58, 60–61
polytheism, 36
postmodernism, 51–54
pseudoscience, 20, 42, 44, 63

R
relativism, 20–21, 53

repressed memory, 59
Russell, Bertrand, 7, 21–22, 26, 40

S
Sagan, Carl, 42–43, 63, 83,
Science for All Americans, 27, 39, 49
scientific creationism, 41
scientific habits of mind, 25, 29, 30, 32, 36, 39–40, 53, 56, 66, 88
Scopes, John, 29, 39
self-criticism in science, 43
shaman, 33, 52, 57
Siegel, Harvey, 50
Smith, Mike
Sokol, Alan, 53
Southerland, Sherry

T
tentativeness in science, 26

U
universal acid, 44

W
Wallace, Alfred, 4, 14–15, 93
Wandersee, James, 9
Wilson, Edward, 44, 93–95, 99
Wolpert, Lewis, 6, 27, 40, 60

Y
young-Earth beliefs, 30, 31, 38, 64, 89

A BOOK SERIES OF CURRICULUM STUDIES

This series employs research completed in various disciplines to construct textbooks that will enable public school teachers to reoccupy a vacated public domain—not simply as "consumers" of knowledge, but as active participants in a "complicated conversation" that they themselves will lead. In drawing promiscuously but critically from various academic disciplines and from popular culture, this series will attempt to create a conceptual montage for the teacher who understands that positionality as aspiring to reconstruct a "public" space. *Complicated Conversation* works to resuscitate the progressive project—an educational project in which self-realization and democratization are inevitably intertwined; its task as the new century begins is nothing less than the intellectual formation of a public sphere in education.

The series editor is:

> Dr. William F. Pinar
> Department of Curriculum and Instruction
> 223 Peabody Hall
> Louisiana State University
> Baton Rouge, LA 70803-4728

To order other books in this series, please contact our Customer Service Department:

> (800) 770-LANG (within the U.S.)
> (212) 647-7706 (outside the U.S.)
> (212) 647-7707 FAX

Or browse online by series:

> www.peterlangusa.com